PREFACE

The world around us is built from atoms and molecules, and the periodic table is our map to this hidden universe. It's not just a chart of numbers and symbols—it's the story of the elements that make up everything we see, touch, and use.

This book, *Interesting Facts About All Elements of Periodic Table*, is my way of bringing that story to life. Drawing from more than a decade of experience in the pharmaceutical industry, I have seen how deeply chemistry shapes our daily lives. That journey, along with my love for science education, inspired me to create a book that is both informative and fun.

From the very first element, Hydrogen, to the heaviest, Oganesson, each element has its own character, history, and importance. In these pages, you'll discover symbols, atomic numbers, masses, electronic configurations, and—most importantly—fascinating facts and stories that make each element memorable.

Through my website *ChemistryNotesInfo* and my podcast *Learn Science and Chemistry*, I have connected with learners across the world. My hope is that this book will spark the same sense of wonder and curiosity in you—because science is not just about facts, it's about discovery.

विज्ञान से जुड़े रहिए!
Stay connected with science!

— जितेन्द्र सिंह संधू

(Jitendra Singh Sandhu)

Index

Hydrogen 1

Helium 3

Lithium 5

Beryllium 7

Boron 9

Carbon 11

Nitrogen 13

Oxygen 15

Fluorine 17

Neon 19

Sodium 21

Magnesium 23

Aluminium 25

Silicon 27

Phosphorus 29

Sulphur 31

Chlorine 33

Argon 35

Potassium 37

Calcium 39

Scandium 41

Titanium 43

Vanadium 45

Chromium 47

Manganese 49

Iron 51

Cobalt 53

Nickel 55

Copper 57

INTERESTING FACTS ABOUT ALL ELEMENTS OF PERIODIC TABLE

JITENDRA SINGH SANDHU

ACKNOWLEDGMENT

This book would not have been possible without the encouragement, guidance, and support I received along the way.

I would like to express my heartfelt gratitude to my family, who stood beside me with patience and love throughout this journey. Their constant motivation gave me the strength to complete this work.

I am deeply thankful to my teachers and colleagues from the pharmaceutical industry, whose knowledge and experience shaped my understanding of science and inspired me to share it with others.

A special thanks to the readers, students, and science enthusiasts who follow *ChemistryNotesInfo* and my podcast *Learn Science and Chemistry*. Your curiosity and feedback continue to inspire me every day.

Finally, I thank every learner who picks up this book. It is written for you—with the hope that it sparks curiosity, builds knowledge, and makes science a little more exciting.

Zinc 59

Gallium 61

Germanium 63

Arsenic 65

Selenium 67

Bromine 69

Krypton 71

Rubidium 73

Strontium 75

Yttrium 77

Zirconium 79

Niobium 81

Molybdenum 83

Technetium 85

Ruthenium 87

Rhodium 89

Palladium 91

Silver 93

Cadmium 95

Indium 97

Tin 99

Antimony 101

Tellurium 103

Iodine 105

Xenon 107

Cesium 109

Barium 111

Lanthanum 113

Cerium 115

Praseodymium 117

Neodymium 119

Promethium...121

Samarium..123

Europium 125

Gadolinium 127

Terbium 129

Dysprosium 131

Holmium 133

Erbium...135

Thulium 137

Ytterbium..139

Lutetium..141

Hafnium 143

Tantalum 145

Tungsten..147

Rhenium...149

Osmium..151

Iridium 153

Platinum 155

Gold 157

Mercury 159

Thallium..161

Lead 163

Bismuth...165

Polonium...167

Astatine 169

Radon...171

Francium...173

Radium...175

Actinium 177

Thorium 179

Protactinium...181

Uranium 183

Neptunium..185

Plutonium...187

Americium...189

Curium...191

Berkelium...193

Californium..195

Einsteinium 197

Fermium 199

Mendelevium 201

Nobelium 203

Lawrencium..205

Rutherfordium 207

Dubnium 209

Seaborgium...211

Bohrium 213

Hassium 215

Meitnerium...217

Darmstadtium...219

Roentgenium...221

Copernicium 223

Nihonium 225

Flerovium..227

Moscovium 229

Livermorium...231

Tennessine 233

Oganesson...235

HYDROGEN

Hello curious minds! I'm Hydrogen, the lightest element and the foundation of the universe. You'll find me in stars, water, and even fuelling the future of energy!

INTERESTING FACTS ABOUT HYDROGEN

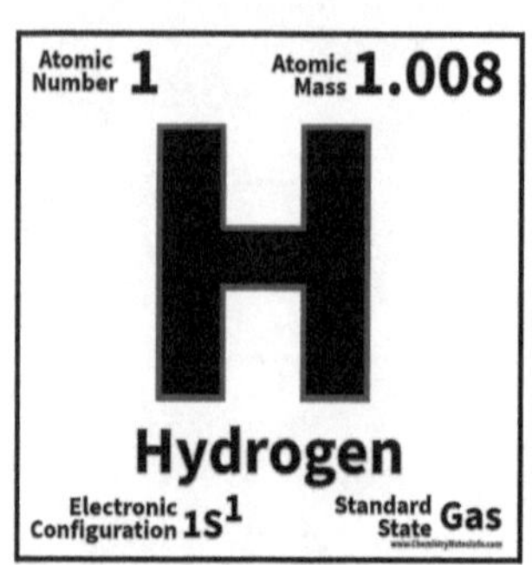

❖ Hydrogen is the most abundant element in the universe. Hydrogen making up about 75% of universe elemental mass.

❖ Hydrogen is the lightest element on the periodic table, with an atomic number of 1. Its atomic weight is approximately about 1.008 atomic mass units.

❖ In its pure gaseous form, hydrogen is colourless, odourless, and tasteless. It is highly flammable when mixed with oxygen.

❖ Hydrogen is generally considered as a clean and sustainable fuel. When hydrogen burns, it produces only water vapour as byproduct, which makes it an environment friendly source of energy. Many big countries like US, UK and India starts utilising it as fuel in real world.

❖ Hydrogen has three isotopes: protium (1H), deuterium (2H), and tritium (3H). Deuterium is widely used in nuclear fusion reactions, and tritium is radioactive and used in some of the nuclear applications.

❖ Hydrogen is known for its ability to form hydrogen bonds with other atoms, especially with oxygen and nitrogen. These bonds are crucial in many biological processes and the structure of water.

❖ In stars, including our Sun, hydrogen undergoes nuclear fusion reaction to form helium. This nuclear fusion reaction releases a tremendous amount of the energy. This energy released during reaction is the source of the sun's energy.

❖ At very low temperatures (around -253°C or -423°F), hydrogen achieves its liquid state. Liquid hydrogen is also used as rocket fuel due to its high energy density.

❖ Storing hydrogen safely and efficiently is a very challenging task. It can be stored as a gas, liquid, or in solid materials. Scientists are working on various technologies development for hydrogen storage. When scientists developed perfect technology for hydrogen storage then it will become more usable in daily life.

❖ Hydrogen is used in various chemical industries. For example, hydrogen is used in industries for the production of ammonia for fertilisers, oil refining industries, and as a reducing agent in metal production industries.

PHYSICAL PROPERTIES:

Physical Property	Description
Atomic Symbol	H
Atomic Number	1
Atomic Mass	1.008 amu
State at Room Temperature	Gas
Colour, Odor, Taste	Colourless, Odourless, Tasteless
Density	0.0899 g/L (at 0°C, 1 atm)
Melting Point	-259.16°C (-434.49°F)
Boiling Point	-252.87°C (-423.17°F)
Density (at 0°C, 1 atm)	0.0899 g/L
Molar Mass	2.01588 g/mol
Hydrogen Bonding	Can form hydrogen bonds with other elements.
Flammability	Highly flammable when mixed with oxygen.
Solubility	Slightly soluble in water.
Abundance in Earth	0.14% by mass, making it the 10th most abundant element in the crust

CHEMICAL PROPERTIES:

Chemical Property	Description
Reactivity	Highly reactive; forms compounds with many elements.
Oxidation States	-1, 0, +1
Electronegativity	2.20
Reaction with Oxygen	$2H_2 + O_2 \rightarrow 2H_2O$ (vigorous exothermic reaction, explosive in air)
Combustion Products	Water ($H2O$) when burned in the presence of oxygen.
Common Compounds	H_2O (Water), HCl (Hydrogen Chloride), NH_3 (Ammonia), CH_4 (Methane)
Reaction with Metals	Forms metal hydrides (e.g., NaH, CaH_2, LiH)
Acid-Base Behavior	Proton donor (H^+ ion); forms strong acids like HCl, H_2SO_4
Isotopes	Protium (1H), Deuterium (2H), Tritium (3H)
Biological Importance	Essential for life; component of water, organic compounds, biomolecules (proteins, DNA, carbohydrates, fats); involved in acid–base balance and cellular energy (ATP)

HELIUM

Hi Science enthusiast! I'm Helium, the party balloon expert and star fuel. Light and carefree, I float through the universe, keeping things cool.

INTERESTING FACTS ABOUT HELIUM

❖ Helium is the second lightest element after hydrogen, with an atomic mass of about 4.0026 atomic mass units.

❖ Helium is a noble gas and is chemically inert, meaning it doesn't easily react with other elements or compounds.

❖ Helium was first discovered in 1868 by astronomers Jules Janssen and Norman Lockyer during a solar eclipse, hence its name, which comes from "Helios," the Greek word for the Sun.

❖ Helium is the second most abundant element in the universe after hydrogen, primarily formed through nuclear fusion in stars.

❖ Unlike hydrogen, helium is non-flammable and is used in airships and balloons as a safer alternative to hydrogen.

❖ Helium exhibits superfluidity at extremely low temperatures. This means it can flow without friction, even climbing walls of containers!

❖ Helium is commonly used to fill balloons due to its low density, which makes the balloons float.

❖ Liquid helium is used in cryogenics, particularly in cooling superconducting magnets, such as those in MRI scanners.

❖ Helium has two stable isotopes: Helium-3 and Helium-4. Helium-4 is by far the most common, making up 99.99986% of naturally occurring helium.

❖ Helium is produced on Earth by the radioactive decay of heavier elements like uranium and thorium.

❖ Helium has the lowest boiling point (-268.93°C) and melting point (-272.2°C) of any element.

❖ Despite its abundance in the universe, helium is relatively rare on Earth and is extracted from natural gas reserves.

PHYSICAL PROPERTIES:

Physical Property	Description
Atomic Symbol	He
Atomic Number	2
Atomic Mass	4.0026 amu
State at Room Temp	Gas
Color	Colorless
Odor	Odorless
Density	0.1786 g/L (at STP)
Melting Point	-272.2°C (only at extremely high pressure)
Boiling Point	-268.93°C
Solubility in Water	Very low solubility
Conductivity	Poor conductor of electricity
Abundance in Earth	Very low in atmosphere (~0.0005% by volume)

CHEMICAL PROPERTIES:

Chemical Property	Description
Reactivity	Inert, does not form compounds under normal conditions
Flammability	Non-flammable
Electronegativity	Not applicable (does not form bonds)
Oxidation States	0 (does not gain or lose electrons easily)
Reaction with Metals	No known reactions with metals
Reaction with Non-metals	No known reactions with non-metals
Bonding	Does not form bonds
Noble Gas	Part of the group 18 noble gases, meaning it has a full outer electron shell, making it highly stable
Cryogenic Use	Used as a coolant in liquid form in extremely low-temperature applications
Isotopes	Helium-3 (rare) and Helium-4 (common), with Helium-4 being used in most applications

LITHIUM

Hey tech lovers! I'm Lithium, your go-to element for powering batteries in smartphones and electric cars. Small but mighty, I keep the world charged!

INTERESTING FACTS ABOUT LITHIUM

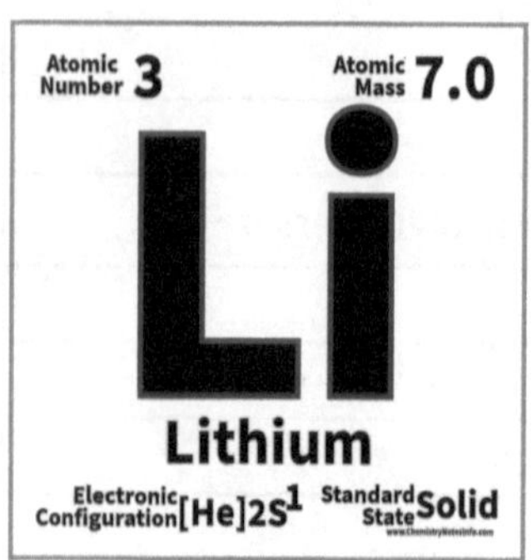

❖ Lithium is the lightest metal and the least dense solid element, with a density about half that of water.

❖ Lithium is highly reactive, especially with water, where it forms lithium hydroxide and hydrogen gas, often accompanied by flames.

❖ Swedish chemist Johan August Arfvedson discovered lithium in the mineral petalite in 1817.

❖ Lithium is a key component in the rechargeable lithium-ion batteries, which are widely used in electronics such as smartphones, laptops, and electric vehicles.

❖ Lithium has a silvery-white appearance, but it tarnishes quickly in air due to oxidation.

❖ Lithium is highly flammable and can ignite in air, making it dangerous to handle in its pure form.

❖ Lithium salts, such as lithium carbonate, are used in psychiatric medicine to treat bipolar disorder and other mood disorders.

❖ Lithium is used in nuclear reactors as a coolant and in nuclear fusion research, where it plays a role in producing tritium.

❖ Lithium is found in various minerals, such as spodumene and lepidolite, as well as in brine deposits, clays, and seawater.

❖ Lithium is the first element in the alkali metal group (Group 1) and shares characteristics with other alkali metals, such as being highly reactive and forming strong bases.

❖ Lithium is added to alloys to improve strength and lightness, commonly used in the aerospace industry to create lightweight materials.

❖ Lithium is also used in rechargeable batteries for smartphones, laptops, and electric vehicles, making it a crucial element in our daily lives.

PHYSICAL PROPERTIES:

Physical Property	Description
Atomic Symbol	Li
Atomic Number	3
Atomic Mass	6.94 amu
State at Room Temp	Solid
Color	Silvery white
Odor	Odorless
Density	0.534 g/cm^3
Melting Point	180.54°C
Boiling Point	1342°C
Solubility in Water	Reacts with water to form lithium hydroxide and hydrogen
Conductivity	Excellent conductor of heat and electricity

CHEMICAL PROPERTIES:

Chemical Property	Description
Reactivity	Highly reactive, especially with water and oxygen
Flammability	Highly flammable, especially when exposed to water or air
Electronegativity	0.98 (Pauling scale)
Oxidation States	1
Reaction with Water	Reacts vigorously with water to produce lithium hydroxide (LiOH) and hydrogen (H_2) gas
Reaction with Oxygen	Forms lithium oxide (Li_2O) and lithium peroxide (Li_2O_2) when burned in air
Reaction with Halogens	Reacts with halogens to form ionic salts like lithium chloride (LiCl)
Hydrides	Forms lithium hydride (LiH) when reacted with hydrogen
Acid-Base Behavior	Forms strong bases like lithium hydroxide when dissolved in water
Alloys	Used in lightweight, strong alloys for the aerospace and automotive industries

BERYLLIUM

Greetings, explorers! I'm Beryllium, known for my strength and lightness. Whether it's in aerospace or gemstones, I'm here to add some sparkle and resilience.

INTERESTING FACTS ABOUT BERYLLIUM

* Beryllium is one of the lightest metals yet extremely strong and stiff, making it valuable in aerospace and military applications.

* Beryllium is transparent to X-rays, which makes it useful in X-ray windows and other medical imaging applications.

* Beryllium and its compounds are highly toxic if inhaled, causing a chronic lung disease known as berylliosis or chronic beryllium disease (CBD).

* Beryllium is named after the mineral beryl $(Be_3Al_2Si_6O_{18})$, which is a source of the element and includes gemstones like emeralds and aquamarines.

* Beryllium was discovered by French chemist Louis-Nicolas Vauquelin in 1798, who found it in beryl and emeralds.

* Beryllium is non-magnetic, which makes it useful in applications requiring non-magnetic materials, such as in electrical components.

* Beryllium has one of the highest melting points among the light metals, making it valuable in high-temperature applications.

* Beryllium is often alloyed with copper to create beryllium copper, a material with excellent electrical conductivity, high strength, and non-sparking properties.

* Beryllium is used in nuclear reactors as a neutron reflector and moderator due to its ability to slow down neutrons.

* Beryl, a mineral containing beryllium, forms beautiful gemstones like emerald and aquamarine due to the presence of trace elements such as chromium and iron.

* Beryllium mirrors are used in space telescopes, like NASA's James Webb Space Telescope, because they remain stable at extremely low temperatures.

* Despite being very strong, beryllium is quite brittle, and it can fracture easily if subjected to impact.

PHYSICAL PROPERTIES:

Physical Property	Description
Atomic Symbol	Be
Atomic Number	4
Atomic Mass	9.0122 amu
State at Room Temp	Solid
Color	Steel-gray
Odor	Odorless
Density	1.85 g/cm^3
Melting Point	1287°C
Boiling Point	2469°C
Solubility in Water	Insoluble
Conductivity	Good conductor of heat and electricity

CHEMICAL PROPERTIES:

Chemical Property	Description
Reactivity	Relatively low reactivity compared to other Group 2 elements (alkaline earth metals)
Oxidation States	2
Reaction with Water	Does not react with water at room temperature, unlike other Group 2 metals
Reaction with Oxygen	Forms a protective oxide layer (BeO) when exposed to air
Reaction with Acids	Reacts with acids, such as hydrochloric acid, to form beryllium chloride ($BeCl_2$) and hydrogen gas
Alloys	Beryllium-copper alloys are used in springs, electrical contacts, and non-sparking tools
Reaction with Halogens	Reacts with halogens to form beryllium halides (e.g., $BeCl_2$)
Bonding	Forms predominantly covalent bonds, unlike other Group 2 elements
Toxicity	Highly toxic if inhaled, with severe long-term health effects (berylliosis)

BORON

Hello innovators! I'm Boron, a master of toughness and heat resistance. From glassmaking to electronics, I've got your back in all things strong and smart.

INTERESTING FACTS ABOUT BORON

❖ Boron is a metalloid, meaning it has properties of both metals and non-metals. It's a brittle, hard substance.

❖ Boron is an essential nutrient for plants, playing a key role in their growth, particularly in cell wall formation and reproductive processes.

❖ Boron is used in a type of cancer treatment called Boron Neutron Capture Therapy (BNCT), which targets cancer cells with precision.

❖ Boron nitride is a compound of boron that is almost as hard as diamond and is used as an abrasive and in cutting tools.

❖ Boron was first isolated by chemists Joseph Louis Gay-Lussac and Louis Jacques Thénard in 1808.

❖ Boron is a key component of borosilicate glass, which is heat-resistant and used in cookware and laboratory equipment (like Pyrex).

❖ Boron is not created in stars through nuclear fusion like many elements. Instead, it is thought to be formed by cosmic ray spallation, where high-energy particles strike larger atomic nuclei.

❖ The isotope boron-10 has a high ability to absorb neutrons, making it useful in nuclear reactors and radiation shielding.

❖ Diborane (B_2H_6), a compound of boron, was once explored as a rocket fuel due to its high energy content, though it was ultimately deemed too reactive.

❖ Pure boron is extremely hard (around 9.5 on the Mohs scale), only slightly softer than diamond.

❖ Boron fibers are incredibly strong yet lightweight, making them useful in aerospace materials and advanced sporting goods.

❖ Boron compounds burn with a distinctive green flame, which makes them useful in fireworks and signal flares.

"

PHYSICAL PROPERTIES:

Physical Property	Description
Atomic Symbol	B
Atomic Number	5
Atomic Mass	10.81 amu
State at Room Temp	Solid
Color	Black-brown
Odor	Odorless
Density	2.34 g/cm^3
Melting Point	2076°C
Boiling Point	3927°C
Solubility in Water	Insoluble
Conductivity	Poor conductor of electricity at room temperature, but it becomes good conductor at high temperature
Abundance in Earth	Low; found mainly in borates (e.g., borax, kernite)

CHEMICAL PROPERTIES:

Chemical Property	Description
Reactivity	Reacts with oxygen to form boron oxide (B_2O_3) when heated, and reacts with halogens to form boron halides
Oxidation States	3
Electronegativity	2.04 (Pauling scale)
Reaction with Acids	Does not react easily with acids at room temperature, but dissolves in concentrated acids at high temperatures
Reaction with Alkalis	Reacts with strong alkalis to form borates
Reaction with Metals	Forms borides with metals (e.g., MgB_2), which have high hardness and melting points
Hydrides	Forms boron hydrides (e.g., diborane, B_2H_6), which are highly reactive
Toxicity	Most boron compounds are not toxic to humans in small amounts, but can be toxic in large quantities
Glass Production	Used to produce borosilicate glass, which is resistant to thermal shock

CARBON

Hi Earth dwellers! I'm Carbon, the backbone of life. From diamonds to DNA, I'm the building block of all living things. Life wouldn't be the same without me!

INTERESTING FACTS ABOUT CARBON

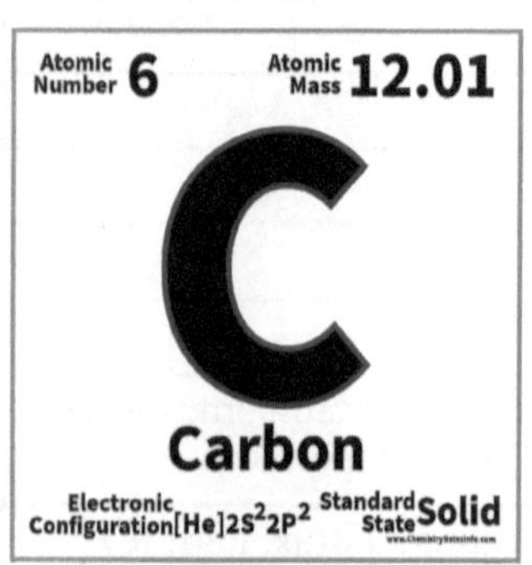

❖Carbon is the backbone of all known life forms. It is present in DNA, proteins, fats, and all organic compounds, making life on Earth carbon-based.

❖Carbon exists in several allotropes, the most well-known being graphite, diamond, and graphene. These forms have vastly different physical properties, with diamond being the hardest naturally occurring material and graphite being soft & slippery.

❖The radioactive isotope carbon-14 is used in radiocarbon dating to estimate the age of archaeological samples, fossils, and historical artifacts.

❖ Carbon dioxide (CO_2) is a critical greenhouse gas and plays a significant role in Earth's climate regulation and in photosynthesis, where plants convert CO_2 into oxygen.

❖ A single layer of carbon atoms arranged in a hexagonal lattice, graphene is one of the strongest materials known and an excellent conductor of heat and electricity.

❖ Diamond, an allotrope of carbon, is the hardest natural substance known and is used in cutting tools, abrasives, and jewellery.

❖ Carbon is the fourth most abundant element in the universe by mass, after hydrogen, helium, and oxygen.

❖ Carbon constantly cycles through the Earth's atmosphere, oceans, and living organisms in the carbon cycle, which is essential for regulating Earth's climate.

❖ Carbon is primary element in fossil fuels like coal, oil, and natural gas. When burned, these fuels release energy and carbon dioxide.

❖ Carbon nanotubes are stronger than steel yet lighter than aluminum, is very important in nanotechnology and materials science.

❖ Carbon, in the form of charcoal, has been used for thousands of years as a fuel and in various chemical processes, such as metal smelting.

PHYSICAL PROPERTIES:

Physical Property	Description
Atomic Symbol	C
Atomic Number	6
Atomic Mass	12.011 amu
State at Room Temp	Solid
Color	Black (graphite), colorless (diamond)
Odor	Odorless
Density	2.267 g/cm³ (graphite), 3.513 g/cm³ (diamond)
Melting Point	Sublimes at 3915°C (graphite)
Boiling Point	4827°C
Solubility in Water	Insoluble
Conductivity	Graphite is a good conductor of electricity; diamond is an electrical insulator
Abundance in Earth	Found in Earth's crust, oceans, and atmosphere; forms the basis of all organic compounds

CHEMICAL PROPERTIES:

Chemical Property	Description
Reactivity	Carbon can react with oxygen to form carbon dioxide (CO_2) or carbon monoxide (CO) when burned
Oxidation States	+4, +2, -4
Electronegativity	2.55 (Pauling scale)
Bonding	Forms covalent bonds with many elements, including hydrogen, oxygen, nitrogen, and other carbons
Hydrocarbons	Carbon forms the backbone of hydrocarbons, which are key in fuels, plastics, and chemical industries
Reaction with Metals	Carbon forms carbides with metals (e.g., calcium carbide, CaC_2), used in various industrial processes
Allotropes	Graphite (soft, lubricating), diamond (hardest natural substance), and fullerenes (carbon spheres)
Reaction with Oxygen	Combusts in oxygen to form CO_2 or CO, depending on oxygen availability
Organic Chemistry	Forms the basis of organic chemistry, which studies carbon-containing compounds
Isotopes	Carbon-12 and Carbon-13 are stable; Carbon-14 is radioactive and used in radiocarbon dating

NITROGEN

Hello atmosphere breathers! I'm Nitrogen, making up most of the air you breathe. From proteins to fertilisers, I'm essential for life on Earth!

INTERESTING FACTS ABOUT NITROGEN

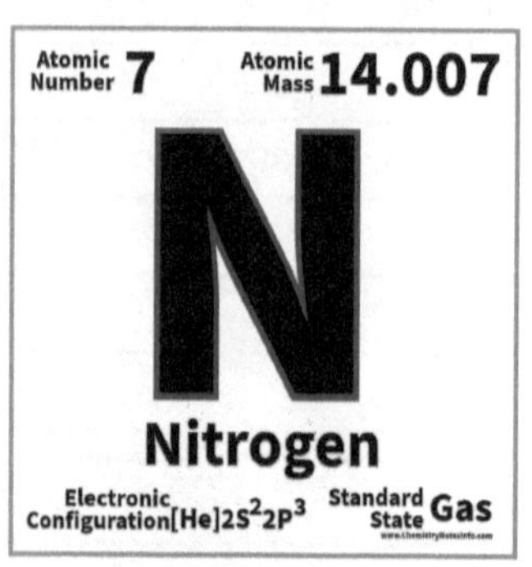

❖ Nitrogen makes up about 78% of the Earth's atmosphere, far more than oxygen, which accounts for about 21%.

❖ Nitrogen is a vital component of amino acids, proteins, and DNA, making it essential for all living organisms.

❖ Nitrogen is relatively inert (non-reactive) at standard conditions, making it useful in preserving food and creating an oxygen-free environment for industrial processes.

❖Nitrogen undergoes a natural cycle in the environment, converting between different chemical forms. Nitrogen-fixing bacteria convert atmospheric nitrogen into a form usable by plants.

❖ Nitrogen becomes a liquid at extremely low temperatures (-196°C) and is used for cryogenic freezing, preserving biological specimens, and in food preparation (like ice cream).

❖ Scottish chemist Daniel Rutherford discovered nitrogen in 1772 when he identified it as "noxious air" that did not support combustion or life.

❖ Nitrogen is used in the Haber process to produce ammonia (NH_3), a key ingredient in fertilisers, explosives, and other chemicals.

❖ Nitrogen compounds such as nitrous oxide (N_2O), also known as laughing gas, are used as anaesthetics and in some performance-enhancing applications like car racing.

❖ Nitrogen is present in many explosive compounds, including TNT and nitroglycerin, due to the instability of nitrogen bonds in such compounds.

❖ Nitrogen Deficiency in Plants: A lack of nitrogen can lead to poor plant growth, as nitrogen is necessary for chlorophyll production and photosynthesis.

PHYSICAL PROPERTIES:

Physical Property	Description
Atomic Symbol	N
Atomic Number	7
Atomic Mass	14.007 amu
State at Room Temp	Gas
Color	Colorless
Odor	Odorless
Density	1.251 g/L (at 0°C and 1 atm)
Melting Point	-210.0°C
Boiling Point	-195.8°C
Solubility in Water	Slightly soluble
Conductivity	Non-conductive
Abundance in Earth	78% of Earth's atmosphere

CHEMICAL PROPERTIES:

Chemical Property	Description
Reactivity	Nitrogen is inert at room temp. but forms compounds with O2, H2, and metals at higher temp.
Oxidation States	-3, +1, +2, +3, +4, +5
Electronegativity	3.04 (Pauling scale)
Reaction with Oxygen	Forms nitrogen oxides (NO_x) when burned in the presence of oxygen at high temperatures
Reaction with Hydrogen	Reacts with hydrogen to form ammonia (NH_3) in the Haber process, a critical reaction for fertilizers
Reaction with Metals	Forms nitrides when reacted with metals at high temperatures, such as magnesium nitride (Mg_3N_2)
Inert in Atmosphere	Pure nitrogen is non-reactive and forms no compounds under normal atmospheric conditions
Nitrogen Oxides	Nitrogen forms various oxides, including nitric oxide (NO) and nitrogen dioxide (NO_2), key components of air pollution
Toxicity	Nitrogen gas (N_2) is non-toxic, but some nitrogen compounds, like nitrogen dioxide (NO_2), are harmful to health
Biological Importance	An essential nutrient in the nitrogen cycle, critical for protein synthesis in all living organisms

OXYGEN

Breathe easy, I'm Oxygen! Found in air and water, I'm the life-supporting element that keeps living beings alive and fires burning bright.

INTERESTING FACTS ABOUT OXYGEN

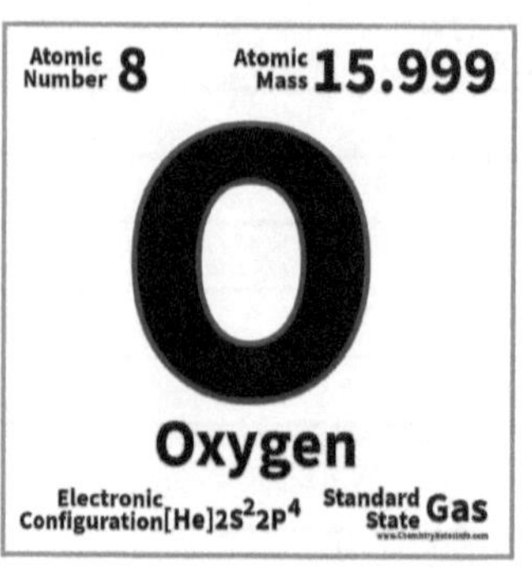

❖ Oxygen makes up about 46% of the Earth's crust by weight, mainly in the form of oxides in minerals.

❖ Oxygen is crucial for respiration in most living organisms, as it is used to produce energy in cells through the process of cellular respiration.

❖ Oxygen, combined with hydrogen, forms water (H_2O), one of the most essential compounds for life on Earth.

❖Oxygen forms ozone (O_3), a molecule composed of three oxygen atoms. The ozone layer in the Earth's atmosphere protects life by absorbing harmful ultraviolet radiation from the Sun.

❖ Oxygen was independently discovered by Carl Wilhelm Scheele and Joseph Priestley in the 1770s, but it was Antoine Lavoisier who named it "oxygen."

❖ Oxygen supports combustion, meaning it is necessary for burning. However, oxygen itself does not burn.

❖ Oxygen is paramagnetic, meaning it is attracted to magnetic fields, a property not common in diatomic molecules.

❖ The oxygen cycle is the natural process by which oxygen is cycled between the atmosphere, biosphere, and lithosphere through processes like photosynthesis and respiration.

❖ Oxygen makes up about 65% of the human body, mostly found in water and organic compounds.

❖ Oxygen becomes a pale blue liquid when cooled to -183°C and is used in rocket fuel as an oxidiser.

❖ In low oxygen environments, such as high altitudes, oxygen deficiency can lead to altitude sickness, and in extreme cases, death.

"""

PHYSICAL PROPERTIES:

Physical Property	Description
Atomic Symbol	O
Atomic Number	8
Atomic Mass	15.999 amu
State at Room Temp	Gas
Color	Colorless
Odor	Odorless
Density	1.429 g/L (at 0°C and 1 atm)
Melting Point	-218.8°C
Boiling Point	-183.0°C
Solubility in Water	Slightly soluble
Conductivity	Non-conductive
Abundance in Earth	46% in crust (by weight); 21% in atmosphere (by volume)

CHEMICAL PROPERTIES:

Chemical Property	Description
Reactivity	Highly reactive, forms compounds with most elements, including oxides, peroxides, and superoxides
Oxidation States	-2, -1 (in peroxides)
Electronegativity	3.44 (Pauling scale)
Reaction with Hydrogen	Reacts with hydrogen to form water (H_2O)
Reaction with Metals	Forms metal oxides, such as iron oxide (rust) or aluminum oxide
Combustion	Oxygen is essential for combustion reactions, combining with carbon, hydrogen, and other fuels to release energy
Oxygen in Air	Makes up 21% of Earth's atmosphere
Allotropes	Exists in two allotropes: dioxygen (O_2) and ozone (O_3)
Toxicity	Oxygen is essential for life but can be toxic at high concentrations or high pressures (oxygen toxicity)
Biological Importance	Critical for aerobic respiration in most living organisms

FLUORINE

Watch out, I'm Fluorine! I might be the most reactive element, but I also keep your teeth strong and your metals corrosion-free.

INTERESTING FACTS ABOUT FLUORINE

❖ Fluorine has the highest electronegativity (3.98 on the Pauling scale) of all elements, meaning it strongly attracts electrons in chemical bonds.

❖ Fluorine is the most reactive of all the elements. It readily forms compounds with almost every other element, including noble gases like xenon and krypton.

❖ In its elemental form, fluorine is a pale yellow, corrosive, and highly toxic gas.

❖ Fluoride, a compound of fluorine, is commonly added to drinking water and toothpaste to help prevent tooth decay.

❖ Hydrofluoric acid is a highly dangerous compound of fluorine that can dissolve glass, making it useful in industrial applications but also extremely hazardous.

❖ Fluorocarbons, such as Teflon (polytetrafluoroethylene), are polymers made with fluorine and are known for their non-stick and heat-resistant properties.

❖ French chemist Henri Moissan isolated fluorine in 1886 after numerous failed and dangerous attempts by previous chemists, earning him the Nobel Prize in Chemistry in 1906.

❖ Despite being inert, noble gases like xenon can form compounds with fluorine, such as xenon hexafluoride (XeF_6), due to fluorine's extreme reactivity.

❖ Fluorine compounds are used in various pharmaceuticals, such as in anaesthetics like halothane and certain antidepressants and antibiotics.

❖ Fluorine is used to produce uranium hexafluoride (UF_6), which is crucial in the nuclear industry for uranium enrichment.

❖ The term "fluorescence" is derived from fluorite (a mineral, CaF_2), which can emit visible light when exposed to ultraviolet light.

PHYSICAL PROPERTIES:

Physical Property	Description
Atomic Symbol	F
Atomic Number	9
Atomic Mass	18.998 amu
State at Room Temp	Gas
Color	Pale yellow
Odor	Pungent, irritating odor
Density	1.696 g/L (at 0°C and 1 atm)
Melting Point	-219.67°C
Boiling Point	-188.11°C
Solubility in Water	Slightly soluble
Conductivity	Non-conductive
Abundance in Earth	Found in minerals like fluorite (CaF_2)

CHEMICAL PROPERTIES:

Chemical Property	Description
Reactivity	Extremely reactive, forms compounds with nearly every other element
Oxidation States	-1 (in most compounds)
Electronegativity	3.98 (Pauling scale)
Reaction with Hydrogen	Reacts violently with hydrogen to form hydrogen fluoride (HF)
Reaction with Metals	Reacts with metals to form fluorides, such as sodium fluoride (NaF) and calcium fluoride (CaF_2)
Fluorine Compounds	Forms strong bonds with other elements, creating compounds like Teflon (polytetrafluoroethylene) and sulfur hexafluoride (SF_6)
Allotropes	Exists only as a diatomic molecule (F_2)
Toxicity	Highly toxic, especially in gas form; hydrofluoric acid can cause severe burns and tissue damage
Biological Importance	Fluorine is not essential to life but is beneficial in small amounts for dental health (fluoride)

NEON

Greetings night owls! I'm Neon, lighting up the night with my glowing, colourful signs. From high voltage to high visibility, I'm all about the glow.

INTERESTING FACTS ABOUT NEON

* Neon is a noble gas, meaning it is chemically inert and rarely forms compounds with other elements due to its complete outer electron shell.

* Neon was discovered by Sir William Ramsay and Morris Travers in 1898 while studying liquefied air. They also discovered krypton and xenon during the same experiments.

* Neon gas is famously used in neon signs. When an electric current passes through the gas, it emits a bright red-orange light.

* Neon is the second lightest noble gas after helium and is about two-thirds as dense as air.

* Although neon is the fifth most abundant element in the universe, it is relatively rare on Earth, making up only about 0.0018% of the atmosphere by volume.

* Neon is non-toxic and poses no threat to human health, which is one reason it is used in various types of lighting.

* Neon is abundant in stars, including our Sun, where it is formed during nuclear fusion processes.

* Despite the term "neon light" being used to describe many coloured lights, pure neon gas glows red-orange. Other gases are used to produce different colours.

* Neon has not been found to form any stable compounds under normal conditions, although some temporary compounds have been created in laboratories.

* Neon is used in cryogenics for refrigerating purposes, and liquid neon is much more efficient than liquid helium in cooling applications.

* Neon is used in helium-neon lasers.

PHYSICAL PROPERTIES:

Physical Property	Description
Atomic Symbol	Ne
Atomic Number	10
Atomic Mass	20.180 amu
State at Room Temp	Gas
Color	Colorless
Odor	Odorless
Density	0.9002 g/L (at 0°C and 1 atm)
Melting Point	-248.59°C
Boiling Point	-246.08°C
Solubility in Water	Insoluble
Conductivity	Non-conductive
Abundance in Earth	0.0018% in the atmosphere

CHEMICAL PROPERTIES:

Chemical Property	Description
Reactivity	Neon is chemically inert and does not react with other elements or compounds under normal conditions
Oxidation States	None (Neon does not form compounds)
Electronegativity	None (Neon does not form bonds)
Reaction with Hydrogen	No reaction with hydrogen
Reaction with Metals	Does not form compounds with metals
Neon in Light	When electrically charged, neon emits a bright red-orange glow, used in neon signs
Allotropes	Neon exists only as a monatomic gas (Ne)
Toxicity	Neon is non-toxic and poses no health risks
Biological Importance	Neon has no known biological role
Isotopes	Three stable isotopes: Neon-20 (90.48%), Neon-21 (0.27%), and Neon-22 (9.25%)

SODIUM

Hello salt lovers! I'm Sodium, found in your favourite table salt. I react with a bang, but in the kitchen, I'm just here to add flavour to your meals.

INTERESTING FACTS ABOUT SODIUM

❖Sodium is a soft, silvery-white metal that reacts violently with water, producing hydrogen gas and heat, which can ignite the gas.

❖Sodium ions (Na$^+$) are vital for nerve transmission, muscle contraction, and maintaining fluid balance in living organisms.

❖Sodium was first isolated by Sir Humphry Davy in 1807 through the electrolysis of molten sodium hydroxide (NaOH).

❖Sodium is most commonly found in nature in the form of sodium chloride (NaCl), or table salt, which is essential for human and animal health.

❖ Sodium vapor lamps are commonly used for street lighting. They produce a distinctive yellow-orange glow.

❖ Due to its high reactivity with water and air, sodium must be stored under mineral oil or in an inert atmosphere to prevent it from reacting with moisture in the air.

❖ Sodium is the sixth most abundant element in the Earth's crust and is found in many minerals, including halite (rock salt), soda ash, and cryolite.

❖ Sodium, along with potassium, helps regulate the body's electrolytes, which are necessary for many physiological functions.

❖ Sodium is used to produce sodium hydroxide (NaOH), a highly caustic substance used in soap-making, cleaning products, and industrial processes.

❖ Sodium compounds such as sodium carbonate (soda ash) are essential in the production of glass.

❖ Sodium is so soft that it can be easily cut with a knife, and it has a melting point of just 97.79°C (207.7°F).

PHYSICAL PROPERTIES:

Physical Property	Description
Atomic Symbol	Na
Atomic Number	11
Atomic Mass	22.990 amu
State at Room Temp	Solid (soft metal)
Color	Silvery-white
Odor	Odorless
Density	0.968 g/cm³ (at 20°C)
Melting Point	97.79°C
Boiling Point	882.9°C
Solubility in Water	Reacts violently
Conductivity	Good conductor of heat and electricity
Abundance in Earth	2.36% in the Earth's crust (by weight)

CHEMICAL PROPERTIES:

Chemical Property	Description
Reactivity	Highly reactive, especially with water and air
Oxidation States	+1 (most common)
Electronegativity	0.93 (Pauling scale)
Reaction with Water	Reacts vigorously with water, producing sodium hydroxide (NaOH) and hydrogen gas (H_2)
Reaction with Oxygen	Reacts with oxygen to form sodium oxide (Na_2O)
Reaction with Halogens	Forms ionic compounds with halogens, such as sodium chloride (NaCl)
Stored in Oil	Must be stored in mineral oil to prevent reaction with air and moisture
Flame Color	Bright yellow when burned, used in flame tests to identify sodium
Toxicity	Elemental sodium is highly reactive and dangerous, but sodium ions (Na^+) are essential for life
Biological Importance	Critical for maintaining fluid balance, nerve function, and muscle contraction in organisms
Isotopes	Sodium-23 is the only stable isotope

MAGNESIUM

Hello health enthusiasts! I'm Magnesium, vital for your bones and muscles. I also burn brightly in fireworks, adding a spark to celebrations.

INTERESTING FACTS ABOUT MAGNESIUM

❖Magnesium is the eighth most abundant element in the Earth's crust and plays a significant role in geology and biology.

❖Magnesium is crucial for many biological processes, including muscle function, nerve transmission, and over 300 enzymatic reactions in the body.

❖Chlorophyll, the green pigment in plants responsible for photosynthesis, contains magnesium at its core, allowing plants to convert sunlight into energy.

❖ Magnesium is a lightweight metal, about one-third lighter than aluminium, and is used in alloys to reduce the weight of structures such as airplanes and cars.

❖ Sir Humphry Davy first isolated pure magnesium in 1808 through electrolysis of a mixture of magnesium oxide and mercury oxide.

❖ When magnesium burns, it produces an intense, bright white flame. It is used in fireworks, flares, and old-fashioned flashbulbs.

❖ Magnesium is the third most abundant element dissolved in seawater, after sodium and chlorine, making it an important component of the ocean's chemistry.

❖ Magnesium is commonly taken as a dietary supplement to support muscle and nerve function, as well as bone health.

❖ Due to its strength and low density, magnesium is used in alloys for automotive parts, aerospace components, and electronics like laptops and cameras.

❖ Magnesium hydroxide, commonly known as milk of magnesia, is used as an antacid and laxative.

❖ Magnesium oxide (MgO) is used in the construction industry for fire-resistant boards, cement, and other materials.

PHYSICAL PROPERTIES:

Physical Property	Description
Atomic Symbol	Mg
Atomic Number	12
Atomic Mass	24.305 amu
State at Room Temp	Solid (metal)
Color	Silvery-white
Odor	Odorless
Density	1.738 g/cm^3
Melting Point	650°C
Boiling Point	1,090°C
Solubility in Water	Insoluble (metal); reacts with water slowly to form magnesium hydroxide ($Mg(OH)_2$)
Conductivity	Good conductor of heat and electricity
Abundance in Earth	2.1% in the Earth's crust

CHEMICAL PROPERTIES:

Chemical Property	Description
Reactivity	Reacts with acids and water to produce hydrogen gas
Oxidation States	+2 (most common)
Electronegativity	1.31 (Pauling scale)
Reaction with oxygen	Forms magnesium oxide(MgO) when exposed to air
Reaction with Water	Reacts slowly with cold water and rapidly with hot water to produce hydrogen gas (H_2) and magnesium hydroxide ($Mg(OH)_2$)
Reaction with Halogens	Reacts with halogens to form halides, such as magnesium chloride ($MgCl_2$)
Flame Color	Burns with an intense white flame, used in fireworks and flares
Toxicity	Magnesium is essential for life, but in high concentrations, magnesium salts can cause toxicity
Biological Importance	Crucial for muscle contraction, nerve function, and enzyme activity

ALUMINIUM

Hi innovators! I'm Aluminium, lightweight and strong. From soda cans to airplanes, I'm the versatile element that's part of your everyday life.

INTERESTING FACTS ABOUT ALUMINIUM

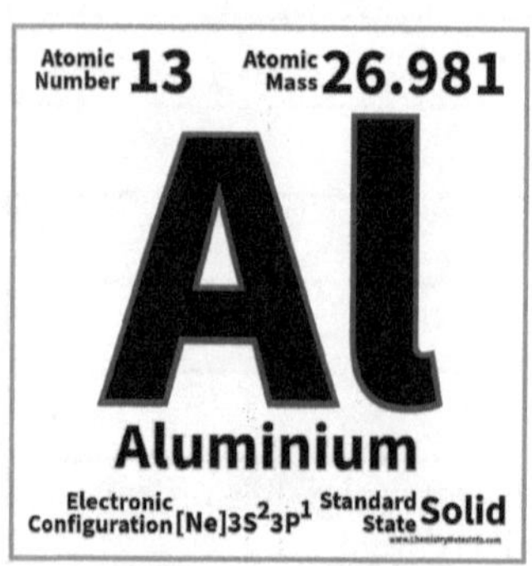

❖Aluminium is the most abundant metal in the Earth's crust, making up about 8.23% of its weight.

❖Aluminium is known for being lightweight and strong, which makes it ideal for use in aerospace, transportation, and construction industries.

❖Aluminium naturally forms a thin oxide layer on its surface when exposed to air, which protects it from corrosion.

❖Hans Christian Ørsted, a Danish chemist, is credited with isolating aluminium in 1825 by reducing aluminium chloride.

❖ Aluminium foil is widely used for food packaging because it is lightweight, non-toxic, and provides a good barrier against moisture, air, and light.

❖ Aluminium is 100% recyclable without losing its properties, and recycling aluminium saves about 95% of the energy required to produce new aluminium.

❖ Due to its high strength-to-weight ratio, aluminium is used in the construction of spacecraft, including the Apollo spacecraft and space stations.

❖ Pure aluminium is soft, but it can be alloyed with elements like copper, magnesium, and silicon to create stronger, more durable materials.

❖ After oxygen and silicon, aluminium is the third most abundant element in the Earth's crust.

❖ Aluminium is primarily obtained from bauxite ore, which contains aluminium oxide (Al_2O_3) and is processed using the Bayer and Hall-Héroult processes.

❖ Aluminium has excellent reflective properties, making it ideal for use in mirrors, solar panels, and decorative items.

PHYSICAL PROPERTIES:

Physical Property	Description
Atomic Symbol	Al
Atomic Number	13
Atomic Mass	26.981 amu
State at Room Temp	Solid (metal)
Color	Silvery-white
Odor	Odorless
Density	2.70 g/cm^3
Melting Point	660.32°C
Boiling Point	2,470°C
Solubility in Water	Insoluble
Conductivity	Good conductor of heat and electricity
Abundance in Earth	8.23% in the Earth's crust

CHEMICAL PROPERTIES:

Chemical Property	Description
Reactivity	Reacts with acids and bases, but is resistant to corrosion due to a protective oxide layer
Oxidation States	+3 (most common)
Electronegativity	1.61 (Pauling scale)
Reaction with Oxygen	Forms aluminum oxide (Al_2O_3), a protective layer that prevents further corrosion
Reaction with Water	At high temperatures, aluminum reacts with water to form aluminum hydroxide ($Al(OH)_3$) and hydrogen gas (H_2)
Reaction with Halogens	Reacts with halogens to form aluminum halides, such as aluminum chloride ($AlCl_3$)
Alloys	Commonly alloyed with elements like copper, magnesium, and zinc to increase strength
Flame Test	Aluminum does not produce a distinctive color in a flame test
Toxicity	Aluminum is not harmful in trace amounts, but high exposure can lead to health issues
Biological Importance	Not essential for life, but it is widely present in food and water
Isotopes	Only one stable isotope: Aluminum-27

SILICON

Hey techies! I'm Silicon, the foundation of the digital world. From microchips to sand, I'm the brainpower behind computers and modern technology.

INTERESTING FACTS ABOUT SILICON

❖Silicon is the second most abundant element in the Earth's crust, making up about 27.7% of its weight, after oxygen.

❖Silicon is the fundamental material used in semiconductors, making it crucial for electronics like computers, smartphones, and solar cells.

❖Silicon combines with oxygen to form silicates, which are the building blocks of most rocks and minerals. Pure silicon dioxide forms quartz.

❖Silicon was first isolated by Swedish chemist Jöns Jacob Berzelius in 1824 by heating potassium with silicon tetrafluoride.

❖ Silicon dioxide (SiO_2) is a primary component of glass. Most glass is made from silica, which is derived from sand.

❖ Silicon is present in trace amounts in the human body and is thought to play a role in the health of bones, skin, and connective tissues.

❖ Silicone, a polymer containing silicon, is used in medical implants, cookware, sealants, lubricants, and many other applications.

❖ Silicon is a key material in photovoltaic cells that convert sunlight into electricity, making it essential for solar energy technologies.

❖ Silicon carbide is an extremely hard material used in abrasives, cutting tools, and semiconductors with high-temperature and high-voltage applications.

❖ Silicon is highly resistant to corrosion and decomposition, which is why silicon-based materials are used in harsh environments like space and deep-sea exploration.

❖ The phrase "Silicon Valley" refers to the region in California known for tech companies and innovations in AI, thanks to the central role of silicon in computing hardware.

PHYSICAL PROPERTIES:

Physical Property	Description
Atomic Symbol	Si
Atomic Number	14
Atomic Mass	28.085 amu
State at Room Temp	Solid (metalloid)
Color	Dark gray with a metallic luster
Odor	Odorless
Density	2.329 g/cm^3
Melting Point	1,414°C
Boiling Point	3,265°C
Solubility in Water	Insoluble
Conductivity	Semiconductor (conductivity can be adjusted by doping with other elements)
Abundance in Earth	27.7% in the Earth's crust

CHEMICAL PROPERTIES:

Chemical Property	Description
Reactivity	Reacts with halogens and certain acids, but is resistant to alkalis and oxygen at low temperature
Oxidation States	+4 (most common), +2
Electronegativity	1.90 (Pauling scale)
Reaction with Oxygen	Reacts with oxygen to form silicon dioxide (SiO_2)
Reaction with Halogens	Reacts with halogens to form silicon halides, such as silicon tetrachloride ($SiCl_4$)
Allotropes	Silicon does not have multiple allotropes, but its crystalline and amorphous forms are significant
Semiconducting Properties	Silicon is a semiconductor, meaning its electrical conductivity can be modified for use in electronics
Flame Test	Silicon does not impart a color to a flame test
Toxicity	Elemental silicon is non-toxic, though fine silicon dust (silica) can be harmful if inhaled
Biological Importance	Present in trace amounts in the body, involved in bone, skin, and connective tissue health
Isotopes	Three stable isotopes: Silicon-28 (92.23%), Silicon-29 (4.67%), Silicon-30 (3.10%)

PHOSPHORUS

Greetings growers! I'm Phosphorus, a key player in fertilisers and DNA. I help plants grow strong and keep cells energised.

INTERESTING FACTS ABOUT PHOSPHORUS

❖Phosphorus was the first element to be discovered in 1669 by German alchemist Hennig Brand, who extracted it from urine while searching for the philosopher's stone.

❖White phosphorus glows with a faint greenish light in the dark due to chemiluminescence, giving phosphorus its name, which means "light-bearer" in Greek.

❖Phosphorus is a critical component of DNA, RNA, and ATP, molecules essential for energy transfer and genetic information in all living organisms.

❖ Phosphorus exists in several forms, or allotropes, including white, red, and black phosphorus. White phosphorus is highly reactive and toxic, while red phosphorus is more stable and is used in safety matches.

❖ Phosphorus is used in safety matches (red phosphorus) and in fireworks to create bright flashes and colours.

❖ Phosphorus is a major component of agricultural fertilisers, necessary for plant growth, often in the form of phosphate compounds like ammonium phosphate.

❖ In the human body, phosphorus is critical for the formation of bones and teeth, where it is present as calcium phosphate.

❖ White phosphorus is used in incendiary bombs and military applications because it ignites spontaneously in air and burns fiercely.

❖ Phosphorus is a nonmetal, yet it is highly reactive, especially in its white form, which must be stored underwater or in inert gases to prevent it from igniting.

❖ White phosphorus is highly toxic and can cause severe burns if it comes into contact with skin. Its fumes are also dangerous if inhaled.

PHYSICAL PROPERTIES:

Physical Property	Description
Atomic Symbol	P
Atomic Number	15
Atomic Mass	30.974 amu
State at Room Temp	Solid (nonmetal)
Color	White (waxy), red, or black, depending on the allotrope
Odor	White phosphorus has a garlic-like smell
Density	White phosphorus: 1.82 g/cm³; Red phosphorus: 2.34 g/cm³
Melting Point	White phosphorus: 44.1°C; Red phosphorus: 590°C (sublimes)
Boiling Point	White phosphorus: 280.5°C
Solubility in Water	Insoluble
Conductivity	Non-conductor of electricity in its common forms
Abundance in Earth	0.099% in the Earth's crust

CHEMICAL PROPERTIES:

Chemical Property	Description
Reactivity	White phosphorus is highly reactive and ignites in air, while red phosphorus is more stable
Oxidation States	-3, +3, +5
Electronegativity	2.19 (Pauling scale)
Reaction with Oxygen	Combines with oxygen to form phosphorus oxides, such as phosphorus pentoxide (P_4O_{10})
Reaction with Water	White phosphorus reacts slowly with water, producing phosphoric acid
Reaction with Halogens	Reacts with halogens to form phosphorus halides, such as phosphorus trichloride (PCl_3)
Flammability	White phosphorus ignites spontaneously in air at temperatures above 30°C
Toxicity	White phosphorus is highly toxic and causes severe burns; red phosphorus is less toxic
Biological Importance	Essential for all living organisms, playing a key role in bones, teeth, and cellular energy production (ATP)

SULPHUR

Hello chemistry lovers! I'm Sulphur, known for my smelly side, but also for making rubber durable and medicines effective.

INTERESTING FACTS ABOUT SULPHUR

❖Sulphur has been known since ancient times, and it was historically referred to as "brimstone." It is mentioned in the Bible and was used in early medicines and alchemy.

❖The characteristic smell of rotten eggs is caused by hydrogen sulphide (H_2S), a sulphur compound released by decaying organic matter.

❖Sulphur is an essential element for all living organisms. It is found in amino acids like cysteine & methionine, which are critical for protein formation.

❖ Large amounts of sulphur are released by volcanic eruptions, and sulphur crystals can be found near volcanic craters and hot springs.

❖ Sulphur is one of the key ingredients in gunpowder, along with charcoal and potassium nitrate.

❖ Sulphur is primarily used in the production of sulphuric acid (H_2SO_4), one of the most important industrial chemicals, used to make fertilisers, batteries, and many other products.

❖ Lead-acid batteries, commonly used in cars, rely on sulphuric acid as an electrolyte to generate electricity.

❖ Sulphur in its pure form is a bright yellow, brittle, and non-metallic solid.

❖ Sulphur exists in several allotropic forms, including rhombic sulphur, monoclinic sulphur, and amorphous sulphur, each with different properties.

❖ Sulphur plays a crucial role in the body's detoxification processes, helping eliminate heavy metals and other harmful substances through sulphur-containing compounds like glutathione.

❖ Sulphur is used in the vulcanisation process of rubber, which makes the rubber more durable and elastic, commonly used in tires.

❖ Sulphur dioxide (SO_2) is used as a preservative in food and wine to prevent oxidation and maintain freshness.

PHYSICAL PROPERTIES:

Physical Property	Description
Atomic Symbol	S
Atomic Number	16
Atomic Mass	32.06 amu
State at Room Temp	Solid (nonmetal)
Color	Yellow (pure form)
Odor	Odorless (elemental form), but sulfur compounds like hydrogen sulfide have a characteristic foul odor
Density	2.07 g/cm³
Melting Point	115.21°C
Boiling Point	444.6°C
Solubility in Water	Insoluble (elemental sulfur); some sulfur compounds are soluble
Conductivity	Non-conductor of electricity
Abundance in Earth	0.042% in the Earth's crust

CHEMICAL PROPERTIES:

Chemical Property	Description
Reactivity	Reacts with many elements, especially oxygen, hydrogen, and halogens
Oxidation States	-2, +4, +6
Electronegativity	2.58 (Pauling scale)
Reaction with Oxygen	Burns in air to form sulfur dioxide (SO_2) and, in the presence of more oxygen, sulfur trioxide (SO_3)
Reaction with Hydrogen	Reacts with hydrogen to form hydrogen sulfide (H_2S), a toxic and flammable gas
Reaction with Metals	Forms metal sulfides when combined with metals, such as iron sulfide (FeS_2)
Flammability	Burns with a blue flame in oxygen
Toxicity	Sulfur itself is non-toxic, but many sulfur compounds (e.g., hydrogen sulfide, sulfur dioxide) are toxic
Biological Importance	Sulfur is vital for amino acids, vitamins, and many enzymes in living organisms
Isotopes	Four stable isotopes: Sulfur-32 (95%), Sulfur-33 (0.75%), Sulfur-34 (4.21%), Sulfur-36 (trace)

CHLORINE

Hi pool-goers! I'm Chlorine, keeping your water clean and safe. From disinfectants to plastics, I'm a vital part of modern living.

INTERESTING FACTS ABOUT CHLORINE

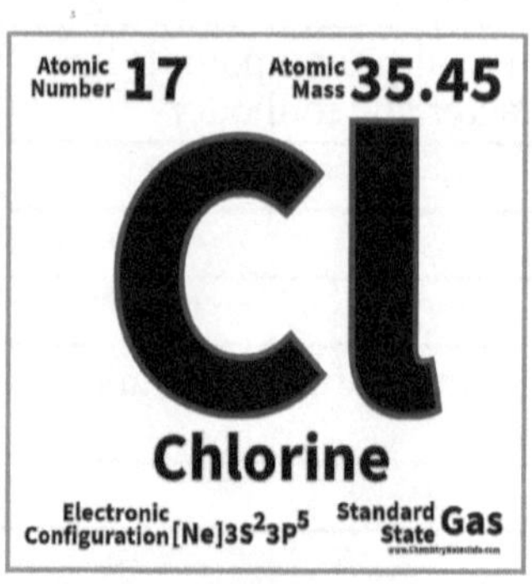

❖Chlorine was first discovered in 1774 by Swedish chemist Carl Wilhelm Scheele, although it was not recognised as an element until 1810 by Sir Humphry Davy.

❖Chlorine is a member of the halogen group (Group 17) and is highly reactive, readily forming compounds with most elements.

❖In its elemental form, chlorine is a greenish-yellow gas that has a strong, pungent odour.

❖Chlorine is widely used as a disinfectant, particularly for purifying drinking water and swimming pools. It kills bacteria, viruses, and other pathogens.

❖ Chlorine combines with sodium to form sodium chloride, or table salt, one of the most common & important compounds in daily life.

❖ Chlorine gas is highly toxic if inhaled, causing respiratory problems and potentially leading to death in high concentrations.

❖ Chlorine is a key component of CFCs, which were once widely used in refrigerants and aerosols but have been phased out due to their role in ozone depletion.

❖ Chlorine compounds, like sodium hypochlorite (NaClO), are used in household bleach and industrial bleaching processes to whiten clothes and disinfect surfaces.

❖ Chloride ions (Cl^-), derived from chlorine, play an essential role in maintaining fluid balance, digestion, nerve function in human body.

❖ Chlorine gas was used as a chemical weapon during World War I, causing severe lung damage and death by suffocation.

❖ Chlorine is often used in organic chemistry for the synthesis of various pharmaceuticals, pesticides, and other chemicals.

❖ Chlorine is abundant in nature, making up about 0.017% of the Earth's crust, and is especially plentiful in seawater in the form of chloride ions.

PHYSICAL PROPERTIES:

Physical Property	Description
Atomic Symbol	Cl
Atomic Number	17
Atomic Mass	35.45 amu
State at Room Temp	Gas
Color	Greenish-yellow
Odor	Strong, pungent odor
Density	3.21 g/L at 0°C (gas); 1.56 g/cm³ (liquid at −34°C)
Melting Point	-101.5°C
Boiling Point	-34.04°C
Solubility in Water	Moderately soluble (forms hypochlorous acid and hydrochloric acid in water)
Conductivity	Non-conductor of electricity as a gas, but ionic compounds like NaCl are good conductors when dissolved or molten
Abundance in Earth	0.017% in the Earth's crust

CHEMICAL PROPERTIES:

Chemical Property	Description
Reactivity	Highly reactive, especially with metals, hydrogen, and organic compounds
Oxidation States	-1 (most common), +1, +3, +5, +7
Electronegativity	3.16 (Pauling scale)
Reaction with Water	Reacts with water to form hydrochloric acid (HCl) and hypochlorous acid (HClO)
Reaction with Hydrogen	Combines explosively with hydrogen to form hydrogen chloride (HCl), which dissolves in water to form hydrochloric acid
Reaction with Metals	Reacts vigorously with most metals to form metal chlorides (e.g., sodium chloride, NaCl)
Flammability	Chlorine is not flammable, but it supports combustion by reacting with other substances
Toxicity	Chlorine gas is highly toxic and can cause severe respiratory damage
Biological Importance	Essential for human body in the form of chloride ions (Cl^-), crucial for maintaining fluid balance & digestion

ARGON

Hello insulators! I'm Argon, the noble gas that keeps things chill. I fill up light bulbs and protect welders, all while staying cool and non-reactive.

INTERESTING FACTS ABOUT ARGON

❖Argon makes up about 0.93% of the Earth's atmosphere, making it the third most abundant gas after nitrogen and oxygen.

❖Argon belongs to the noble gas group (Group 18) and is chemically inert, meaning it doesn't easily react with other elements or compounds.

❖Argon was discovered in 1894 by Sir William Ramsay and Lord Rayleigh while investigating why the density of nitrogen extracted from air differed from nitrogen produced chemically.

❖ Argon is commonly used in incandescent & fluorescent light bulbs to prevent the oxidation of the filament, extending the bulb's life.

❖ Argon is used in lasers that produce a distinctive blue-green light, which is used in various medical procedures, such as eye surgeries.

❖ Argon is used as a shielding gas in welding to protect hot metals from reacting with atmospheric gases like oxygen and nitrogen.

❖ Argon is colorless, odorless, and tasteless in both its gaseous and liquid forms, making it undetectable without special equipment.

❖ Argon produces a pale blue or violet light when electrically excited, and it's often used in neon signs for this purpose.

❖ Argon is not toxic, and it has no known biological role in humans or any other living organisms.

❖ The name "argon" is derived from the Greek word "argos," meaning "inactive" or "lazy," reflecting its chemically inert nature.

❖ Argon is used between the panes of double-glazed windows as an insulator, helping to improve energy efficiency in homes/buildings.

❖ Argon is produced in the cores of stars during stellar nucleosynthesis and is present in the universe, though it's more abundant on Earth due to its stability and inability to escape into space.

PHYSICAL PROPERTIES:

Physical Property	Description
Atomic Symbol	Ar
Atomic Number	18
Atomic Mass	39.948 amu
State at Room Temp	Gas
Color	Colorless
Odor	Odorless
Density	1.784 g/L (at 0°C and 1 atm)
Melting Point	-189.35°C
Boiling Point	-185.85°C
Solubility in Water	Slightly soluble
Conductivity	Non-conductive as a gas (inert)
Abundance in Earth	0.93% in the atmosphere

CHEMICAL PROPERTIES:

Chemical Property	Description
Reactivity	Argon is chemically inert and forms no known stable compounds under normal conditions
Oxidation States	0 (no typical oxidation state since it doesn't form compounds)
Electronegativity	Not applicable (as it doesn't form bonds easily)
Reaction with Oxygen	Argon does not react with oxygen
Reaction with Water	No reaction with water
Reaction with Metals	Does not form compounds with metals
Inertness	Argon is one of the least reactive elements due to its full outer electron shell
Flammability	Non-flammable
Toxicity	Non-toxic, but it can displace oxygen in closed spaces, leading to asphyxiation
Biological Importance	Argon has no known biological role
Isotopes	Three stable isotopes: Argon-36 (0.337%), Argon-38 (0.063%), Argon-40 (99.6%)

POTASSIUM

Hello health buffs! I'm Potassium, essential for keeping your muscles and nerves working right. You'll find me in bananas and your body's cells!

INTERESTING FACTS ABOUT POTASSIUM

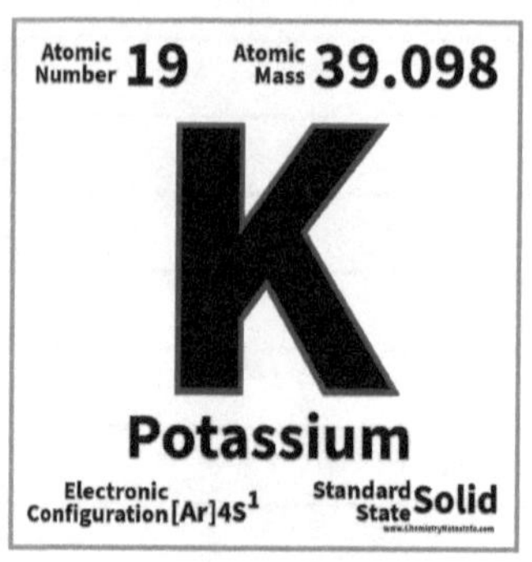

❖The name "potassium" is derived from "potash," a substance obtained by burning wood and leaching the ashes in water. It was first isolated by Sir Humphry Davy in 1807 through electrolysis.

❖Potassium is an essential electrolyte in the human body, helping regulate fluid balance, nerve signals, and muscle contractions, including the heart.

❖Potassium is an extremely reactive metal, especially in water, where it reacts violently, producing hydrogen gas and heat.

❖ In its pure form, potassium is a soft, silvery-white metal that can be cut with a knife. However, it tarnishes quickly when exposed to air due to its high reactivity.

❖ Potassium is famously associated with bananas, although many other foods like avocados, spinach, and potatoes also contain high amounts of potassium.

❖ Potassium is one of the three primary nutrients in fertilisers (alongside nitrogen and phosphorus) and is essential for plant health and crop yields.

❖ Potassium ions (K^+) play a key role in nerve transmission and muscle contraction by helping generate electrical signals in the body's cells.

❖ When potassium is burned, it produces a characteristic lilac or light purple flame, making it useful in flame tests to identify its presence.

❖ Potassium carbonate, also known as potash, is used in the production of glass, soaps, and detergents.

❖ Potassium-40, a naturally occurring isotope, is slightly radioactive and contributes to the natural radioactivity of the human body and Earth's crust.

PHYSICAL PROPERTIES:

Physical Property	Description
Atomic Symbol	K
Atomic Number	19
Atomic Mass	39.0983 amu
State at Room Temp	Solid (metal)
Color	Silvery-white (quickly tarnishes to dull gray)
Odor	Odorless
Density	0.862 g/cm^3 (less dense than water)
Melting Point	63.5°C
Boiling Point	759°C
Solubility in Water	Reacts violently with water, producing potassium hydroxide (KOH) and hydrogen gas
Conductivity	Excellent conductor of electricity
Abundance in Earth	2.6% in the Earth's crust

CHEMICAL PROPERTIES:

Chemical Property	Description
Reactivity	Highly reactive, especially with water and air (oxidizes rapidly in air)
Oxidation States	+1 (most common)
Electronegativity	0.82 (Pauling scale)
Reaction with Water	Reacts explosively with water to form potassium hydroxide (KOH) and hydrogen gas (H_2)
Reaction with Oxygen	Combines with oxygen to form potassium oxide (K_2O) or potassium peroxide (K_2O_2)
Reaction with Halogens	Reacts readily with halogens like chlorine to form potassium halides (e.g., KCl)
Flammability	Highly flammable, ignites spontaneously in air under certain conditions
Toxicity	Potassium compounds like potassium chloride (KCl) are non-toxic in moderate amounts, but imbalances can be dangerous
Biological Importance	Essential for cellular function, nerve transmission, and heart rhythm

CALCIUM

Hi strong bones! I'm Calcium, essential for building strong teeth and bones. From milk to concrete, I make things solid and stable.

INTERESTING FACTS ABOUT CALCIUM

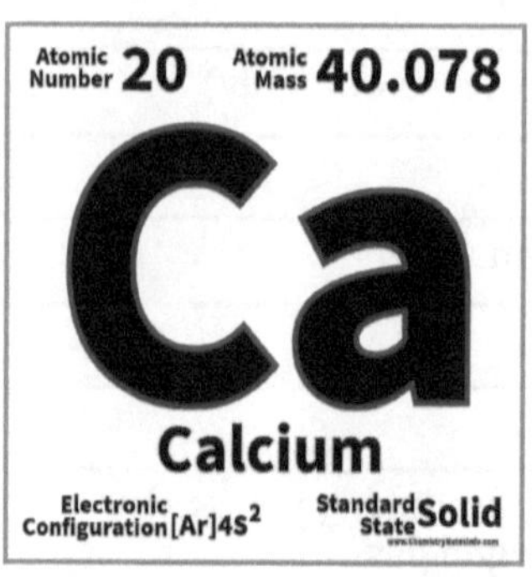

❖Calcium is the most abundant metal in the human body, making up about 1.5% of total body weight, mostly stored in bones and teeth.

❖Calcium is essential for building and maintaining strong bones and teeth. It works with phosphorus to form hydroxyapatite, a key component of bone structure.

❖Calcium was first isolated in 1808 by British chemist Sir Humphry Davy through electrolysis of calcium oxide (lime).

❖ Calcium ions (Ca^{2+}) play a key role in muscle contraction, including the functioning of the heart muscle.

❖ Found in limestone, chalk, and marble, calcium carbonate ($CaCO_3$) is one of the most common calcium compounds and is used in cement, building materials, and antacids.

❖ "Hard water" contains high levels of calcium and magnesium ions. Water hardness can lead to scale buildup in pipes and reduce the effectiveness of soap.

❖ Calcium is essential for blood clotting (coagulation), helping to activate various proteins in the clotting process.

❖ Calcium is the fifth most abundant element in the Earth's crust, making up about 3.6%, mostly found in minerals like limestone and gypsum.

❖ Dairy products like milk, cheese, and yogurt are excellent dietary sources of calcium, crucial for bone development and health.

❖ Calcium compounds, particularly calcium oxide (lime), are used to improve soil quality in agriculture by neutralising acidic soils.

❖ When calcium is burned, it produces a characteristic orange-red flame, making it useful for flame tests in chemistry.

❖ A lack of calcium in the diet can lead to osteoporosis (weak bones).

PHYSICAL PROPERTIES:

Physical Property	Description
Atomic Symbol	Ca
Atomic Number	20
Atomic Mass	40.078 amu
State at Room Temp	Solid (metal)
Color	Silvery-white
Odor	Odorless
Density	1.55 g/cm³
Melting Point	842°C
Boiling Point	1,484°C
Solubility in Water	Slightly soluble as calcium hydroxide ($Ca(OH)_2$)
Conductivity	Good conductor of electricity and heat
Abundance in Earth	3.6% of the Earth's crust

CHEMICAL PROPERTIES:

Chemical Property	Description
Reactivity	Reacts readily with water and acids to form calcium hydroxide and hydrogen gas
Oxidation States	+2 (most common)
Electronegativity	1.00 (Pauling scale)
Reaction with Water	Reacts with cold water to form calcium hydroxide ($Ca(OH)_2$) and hydrogen (H_2) gas
Reaction with Oxygen	Reacts with oxygen to form calcium oxide (CaO), also known as quicklime
Reaction with Halogens	Reacts readily with halogens like chlorine to form calcium halides (e.g., calcium chloride, $CaCl_2$)
Flammability	Calcium itself is not highly flammable but can burn in air under high heat
Toxicity	Non-toxic in small quantities, essential for human health
Biological Importance	Critical for bone health, muscle contraction, nerve signaling, and blood clotting
Isotopes	Six stable isotopes, with Calcium-40 being the most abundant (97%)

SCANDIUM

Hello high-tech enthusiasts! I'm Scandium, a rare metal used in aerospace and sports equipment. I bring lightweight strength to high-performance materials.

INTERESTING FACTS ABOUT SCANDIUM

❖Scandium was discovered in 1879 by Swedish chemist Lars Fredrik Nilson while analyzing the mineral euxenite. It was predicted earlier by Dmitri Mendeleev, who named the element "eka-boron" before its discovery.

❖Scandium is named after the region of Scandinavia, where it was first discovered in minerals.

❖Although it is classified as a rare earth element, scandium is more common than lead or mercury but is widely dispersed in nature, making it difficult to extract in significant quantities.

❖ Scandium is commonly used in aluminum alloys to improve their strength, lightness, and corrosion resistance. These alloys are used in aerospace components, sports equipment, and military applications.

❖ High-performance sports equipment, such as baseball bats and bicycle frames, often use scandium-aluminum alloys for their lightweight yet durable properties.

❖ Scandium's chemical behaviour is similar to that of aluminum and yttrium, which is why it's placed between them in periodic table.

❖ Scandium iodide is used in high-intensity discharge lamps, particularly in stadium lighting, due to its ability to produce a bright, white light that mimics natural sunlight.

❖ Scandium is one of the least abundant elements in the Earth's crust, with an average concentration of only 22 parts per million (ppm).

❖ Scandium is considered non-toxic and has no known biological role, but it is usually found in trace amounts in the human body due to environmental exposure.

❖ Scandium's strength and lightness make it ideal for spacecraft and satellite components.

PHYSICAL PROPERTIES:

Physical Property	Description
Atomic Symbol	Sc
Atomic Number	21
Atomic Mass	44.955908 amu
State at Room Temp	Solid (metal)
Color	Silvery-white
Odor	Odorless
Density	2.985 g/cm^3
Melting Point	1,541°C
Boiling Point	2,831°C
Solubility in Water	Insoluble in water, but scandium salts can dissolve
Conductivity	Good conductor of electricity and heat
Abundance in Earth	22 ppm in the Earth's crust

CHEMICAL PROPERTIES:

Chemical Property	Description
Reactivity	Scandium reacts slowly with oxygen and forms a protective oxide layer (Sc_2O_3) on its surface
Oxidation States	+3 (most common)
Electronegativity	1.36 (Pauling scale)
Reaction with Water	Scandium reacts with water slowly at room temperature, but more rapidly when heated
Reaction with Acids	Scandium dissolves in acids, forming scandium salts (e.g., scandium chloride, $ScCl_3$)
Reaction with Oxygen	Forms a thin layer of scandium oxide (Sc_2O_3) when exposed to air
Reaction with Halogens	Reacts with halogens to form scandium halides, such as scandium fluoride (ScF_3)
Flammability	Not flammable
Toxicity	Non-toxic, with no known biological role
Biological Importance	Scandium has no known essential role in biological systems
Isotopes	Only one naturally occurring stable isotope: Scandium-45

TITANIUM

Greetings adventurers! I'm Titanium, known for being incredibly strong and corrosion-resistant. You'll find me in spacecraft, implants, and even luxury watches.

INTERESTING FACTS ABOUT TITANIUM

❖Titanium was named after the Titans of Greek mythology, reflecting its strength and durability.

❖Titanium was discovered by British clergyman and mineralogist William Gregor in 1791 in a mineral called ilmenite. It was later named "titanium" by German chemist Martin Heinrich Klaproth.

❖Titanium is renowned for its excellent strength-to-weight ratio. It's as strong as steel but only about 45% of the weight, making it ideal for aerospace applications.

❖ Titanium is extremely resistant to corrosion, especially from seawater and chlorine, which is why it's used in marine and chemical industries.

❖ Titanium is non-toxic and biocompatible, meaning it doesn't react with the human body, making it perfect for medical implants like artificial joints, dental implants, and bone plates.

❖ Due to its strength and lightness, titanium is widely used in the aerospace industry for building aircraft frames, engines, and spacecraft components.

❖ Titanium dioxide is widely used as a white pigment in paints, sunscreen, cosmetics, and food products due to its ability to reflect light and provide UV protection.

❖ Titanium is the ninth most abundant element in Earth's crust, making up about 0.6%, but it's challenging to extract in pure form due to its strong bonding with other elements.

❖ Titanium alloys are used in high-performance sports equipment, such as golf clubs, tennis rackets, and bicycle frames, where strength and lightness are crucial.

❖ Titanium is non-magnetic, making it useful in applications that require materials that won't be affected by magnetic fields.

PHYSICAL PROPERTIES:

Physical Property	Description
Atomic Symbol	Ti
Atomic Number	22
Atomic Mass	47.867 amu
State at Room Temp	Solid (metal)
Color	Silvery-gray
Odor	Odorless
Density	4.54 g/cm³
Melting Point	1,668°C
Boiling Point	3,287°C
Solubility in Water	Insoluble
Conductivity	Good conductor of electricity and heat
Abundance in Earth	0.6% in Earth's crust

CHEMICAL PROPERTIES:

Chemical Property	Description
Reactivity	Titanium is reactive with oxygen, forming a protective oxide layer (TiO_2), making it highly corrosion-resistant
Oxidation States	+4 (most common), +3
Electronegativity	1.54 (Pauling scale)
Reaction with Oxygen	Reacts with oxygen to form titanium dioxide (TiO_2), a protective layer that prevents further corrosion
Reaction with Water	Titanium does not react with water at room temperature
Reaction with Acids	Reacts slowly with hydrochloric acid but more rapidly with sulfuric and hydrofluoric acids
Reaction with Halogens	Reacts with halogens like chlorine to form titanium halides, such as titanium tetrachloride ($TiCl_4$)
Flammability	Finely divided titanium powder is flammable, but bulk titanium is non-flammable
Toxicity	Non-toxic and biocompatible; safe for use in medical implants
Biological Importance	No biological role, but widely used in medical implants due to its biocompatibility

VANADIUM

Hi steelworkers! I'm Vanadium, making steel tough and resilient. I help build skyscrapers, bridges, and tools that last a lifetime.

INTERESTING FACTS ABOUT VANADIUM

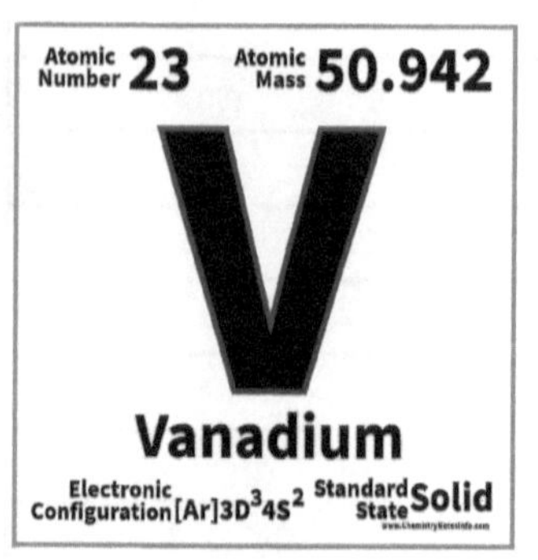

❖Vanadium was first discovered by Spanish-Mexican scientist Andrés Manuel del Río in 1801, but it was mistakenly identified as chromium. It was rediscovered in 1830 by Swedish chemist Nils Gabriel Sefström, who named it vanadium after Vanadis, the Norse goddess of beauty.

❖Vanadium is commonly used in steel alloys to improve strength, durability, & resistance to corrosion, making it essential in tools, knives, & structural materials like rebar.

❖Vanadium is used in vanadium redox flow batteries (VRFB), an emerging technology for large-scale energy storage, particularly for renewable energy like wind and solar power.

❖ Vanadium is often found in magnetite (an iron ore), making vanadium-rich magnetite a source for extracting the element.

❖ Vanadium alloys are extremely corrosion-resistant, which is why they are used in the construction of pipes and other industrial applications exposed to harsh environments.

❖ Due to its high strength and low density, vanadium is used in spacecraft and jet engines where light, strong materials are essential for performance and fuel efficiency.

❖ Vanadium compounds, particularly vanadium pentoxide (V_2O_5), are used in ceramics and glassmaking to produce brilliant colours, such as yellow and green.

❖ Vanadium compounds can exhibit a variety of oxidation states, each displaying different colours, which makes them useful in chemical tests and reactions.

❖ Vanadium is used in high-temperature components of jet engines because of its ability to withstand heat without corroding or degrading.

❖ Vanadium is used as a catalyst in the chemical industry, particularly in the production of sulphuric acid.

PHYSICAL PROPERTIES:

Physical Property	Description
Atomic Symbol	V
Atomic Number	23
Atomic Mass	50.9415 amu
State at Room Temp	Solid (metal)
Color	Silvery-gray
Odor	Odorless
Density	6.11 g/cm^3
Melting Point	1,910°C
Boiling Point	3,407°C
Solubility in Water	Insoluble in water
Conductivity	Good conductor of heat and electricity
Abundance in Earth	About 160 ppm in the Earth's crust

CHEMICAL PROPERTIES:

Chemical Property	Description
Reactivity	Vanadium is stable in air at room temperature due to a thin oxide layer that forms on its surface
Oxidation States	+2, +3, +4, +5 (most common is +5)
Electronegativity	1.63 (Pauling scale)
Reaction with Oxygen	Reacts with oxygen to form vanadium oxides, such as vanadium pentoxide (V_2O_5)
Reaction with Acids	Dissolves in acids, forming vanadium salts (e.g., vanadium chloride, VCl_3)
Reaction with Water	Does not react with water at room temperature
Reaction with Halogens	Reacts with halogens to form vanadium halides, such as vanadium trifluoride (VF_3)
Flammability	Vanadium powder is flammable, but bulk vanadium is non-flammable
Toxicity	Vanadium compounds can be toxic in large amounts
Biological Importance	No essential role in humans, but may have a minor role in regulating insulin and cholesterol levels
Isotopes	Two natural isotopes: Vanadium-50 (0.24%) and Vanadium-51 (99.76%)

CHROMIUM

Hello shiny lovers! I'm Chromium, giving stainless steel its shine and resistance to rust. From chrome finishes to colourful pigments, I'm all about the sparkle.

INTERESTING FACTS ABOUT CHROMIUM

❖The name "chromium" comes from the Greek word "chroma," meaning colour, because chromium compounds are known for their vibrant colours, ranging from yellow to green to red.

❖Chromium was discovered by French chemist Louis-Nicolas Vauquelin in 1797, who found it in the mineral crocoite (lead chromate).

❖Chromium is a key ingredient in stainless steel, providing corrosion resistance and strength. Typically, stainless steel contains about 10.5% chromium.

❖ Chromium is widely used in chrome plating to give a shiny, corrosion-resistant surface to metal objects like car parts, kitchen fixtures, and tools.

❖ Chromium forms various oxides, including chromium(III) oxide (Cr_2O_3), which is used as a green pigment in paints and ceramics.

❖ Chromium is a very hard, brittle metal with high resistance to tarnishing and corrosion. This makes it ideal for use in cutting tools and armour plating.

❖ Chromium, in trace amounts, is essential for human health. It plays a role in regulating blood sugar by enhancing the action of insulin.

❖ The red colour in rubies is caused by trace amounts of chromium impurities in the aluminium oxide (corundum) crystal structure.

❖ Chromium has a high melting point of 1,907°C, making it useful in high-temperature applications like furnace linings and moulds for casting metals.

❖ Chromium compounds, especially lead chromate ($PbCrO_4$), have been used historically in yellow pigments for paints, though many of these are now restricted due to toxicity.

❖ Chromium compounds are used as catalysts in several reactions.

PHYSICAL PROPERTIES:

Physical Property	Description
Atomic Symbol	Cr
Atomic Number	24
Atomic Mass	51.9961 amu
State at Room Temp	Solid (metal)
Color	Silvery-gray with a shiny surface
Odor	Odorless
Density	7.19 g/cm³
Melting Point	1,907°C
Boiling Point	2,671°C
Solubility in Water	Insoluble
Conductivity	Good conductor of electricity and heat
Abundance in Earth	About 100 ppm in the Earth's crust

CHEMICAL PROPERTIES:

Chemical Property	Description
Reactivity	Chromium forms a protective oxide layer, making it resistant to corrosion
Oxidation States	+2, +3, +6 (most common: +3, +6)
Electronegativity	1.66 (Pauling scale)
Reaction with Oxygen	Reacts with oxygen to form chromium oxides (e.g., Cr_2O_3, CrO_3)
Reaction with Acids	Dissolves in strong acids, forming chromium salts (e.g., chromium chloride, $CrCl_3$)
Reaction with Water	Chromium metal does not react with water at room temperature
Reaction with Halogens	Reacts with halogens to form chromium halides, such as chromium fluoride (CrF_3)
Flammability	Chromium powder is flammable, but bulk chromium is non-flammable
Toxicity	Hexavalent chromium (Cr^{6+}) is highly toxic and carcinogenic; trivalent chromium (Cr^{3+}) is essential for health in trace amounts
Biological Importance	Required in trace amounts for regulating blood sugar and insulin function

MANGANESE

Hello builders! I'm Manganese, strengthening steel and helping enzymes work in your body. From construction to biology, I play an essential role.

INTERESTING FACTS ABOUT MANGANESE

❖Manganese was discovered by Swedish chemist Johan Gottlieb Gahn in 1774 when he isolated it from the mineral pyrolusite (manganese dioxide).

❖Manganese is critical in the steel industry, where it is used to remove impurities like sulphur and oxygen from molten steel, improving its strength and durability.

❖Manganese is essential for all living organisms. It plays a key role in the metabolism of carbohydrates, proteins, and fats, as well as in bone formation and antioxidant defence.

❖ Manganese compounds are used as pigments. Manganese dioxide (MnO_2) is used to produce brown pigments, and other manganese compounds can create shades of pink, purple, and green.

❖ Manganese is the 12th most abundant element in the Earth's crust, where it occurs mainly in the form of oxides and silicates.

❖ Vast amounts of manganese exist on the ocean floor in the form of manganese nodules. These nodules contain valuable metals like nickel, copper, and cobalt, making them a potential resource for deep-sea mining.

❖ Manganese is vital in the process of photosynthesis in plants, where it aids in the water-splitting reaction, helping plants convert sunlight into energy.

❖ Manganese dioxide was historically used to remove the green tint from glass caused by iron impurities. It can also produce purple glass when used in larger quantities.

❖ Critical in Alloys: Manganese is used in various alloys, including aluminium alloys, where it improves resistance to corrosion, and in high-strength steel alloys for construction and transportation.

❖ Although the United States is a large consumer of manganese, it has no significant domestic reserves and relies heavily on imports.

PHYSICAL PROPERTIES:

Physical Property	Description
Atomic Symbol	Mn
Atomic Number	25
Atomic Mass	54.938044 amu
State at Room Temp	Solid (metal)
Color	Silvery-gray
Odor	Odorless
Density	7.21 g/cm^3
Melting Point	1,246°C
Boiling Point	2,061°C
Solubility in Water	Insoluble
Conductivity	Good conductor of heat and electricity
Abundance in Earth	About 950 ppm in Earth's crust

CHEMICAL PROPERTIES:

Chemical Property	Description
Reactivity	Manganese reacts slowly with oxygen in air to form manganese oxide
Oxidation States	+2, +3, +4, +6, +7 (most common: +2 and +7)
Electronegativity	1.55 (Pauling scale)
Reaction with Oxygen	Reacts with oxygen to form manganese oxides, such as MnO and MnO_2
Reaction with Acids	Reacts with acids, forming manganese salts (e.g., manganese sulfate, $MnSO_4$)
Reaction with Water	Manganese metal does not react with water at room temperature
Reaction with Halogens	Reacts with halogens to form manganese halides, such as manganese chloride ($MnCl_2$)
Flammability	Manganese powder is flammable, but bulk manganese is non-flammable
Toxicity	Manganese compounds can be toxic in high concentration
Biological Importance	Essential trace element for humans; important for bone health, metabolism, and enzyme function
Isotopes	One stable isotope: Manganese-55

IRON

Greetings, industrial giants! I'm Iron, the backbone of infrastructure. From bridges to your blood, I'm the strong and dependable element that keeps things moving.

INTERESTING FACTS ABOUT IRON

❖Iron is the fourth most abundant element in the Earth's crust and makes up about 5% of it. It is also a key component of the Earth's core.

❖Iron is a critical component of haemoglobin, the protein in red blood cells that transports oxygen throughout the body. Without enough iron, people can develop anaemia.

❖Iron has been used for over 5,000 years, with the Iron Age marking a major advancement in tools and weaponry, transitioning from bronze to stronger iron materials.

❖ The majority of iron mined today is used to produce steel, an alloy of iron and carbon that is essential for construction, transportation, and manufacturing industries.

❖ Iron is magnetic, making it useful in electrical applications such as transformers, motors, and inductors. It is one of only three naturally magnetic elements (the others are cobalt and nickel).

❖ When exposed to water and oxygen, iron forms iron oxide (rust), which is reddish-brown and flaky. Rust weakens iron and can cause structural damage over time.

❖ Iron is an essential dietary mineral, found in foods like red meat, lentils, spinach, and fortified cereals. It plays a key role in energy production and oxygen transport in the body.

❖ Iron can be processed into various forms, including wrought iron, which is tough and malleable, and cast iron, which is hard and brittle. Both have been used historically in construction and tools.

❖ Iron production was a driving force behind the Industrial Revolution in the 18th and 19th centuries, especially with the development of the Bessemer process for making steel.

❖ Iron oxides are used as pigments in paints, ceramics, and cosmetics.

PHYSICAL PROPERTIES:

Physical Property	Description
Atomic Symbol	Fe (from the Latin word "ferrum")
Atomic Number	26
Atomic Mass	55.845 amu
State at Room Temp	Solid (metal)
Color	Lustrous metallic gray
Odor	Odorless
Density	7.87 g/cm^3
Melting Point	1,538°C
Boiling Point	2,862°C
Solubility in Water	Insoluble
Conductivity	Excellent conductor of heat and electricity
Abundance in Earth	Makes up 5% of Earth's crust; abundant in the core

CHEMICAL PROPERTIES:

Chemical Property	Description
Reactivity	Iron readily reacts with oxygen and water to form rust (iron oxide)
Oxidation States	+2, +3 (most common), +4, +6
Electronegativity	1.83 (Pauling scale)
Reaction with Oxygen	Forms iron oxides (e.g., Fe_2O_3, Fe_3O_4) when exposed to air, especially in the presence of moisture
Reaction with Acids	Dissolves in acids, such as hydrochloric acid, producing hydrogen gas and iron salts (e.g., $FeCl_2$)
Reaction with Water	Reacts with water and oxygen over time, leading to rust formation
Reaction with Halogens	Reacts with halogens to form iron halides (e.g., $FeCl_3$, $FeBr_2$)
Flammability	Iron powder is flammable, but bulk iron is not flammable
Toxicity	Iron is non-toxic in moderate amounts, but excessive amounts can cause iron poisoning
Biological Importance	Essential for human health, playing a key role in oxygen transport and enzyme function

COBALT

Hello explorers! I'm Cobalt, colouring the world blue and powering your rechargeable batteries. You'll find me in paints, electronics, and even space exploration.

INTERESTING FACTS ABOUT COBALT

❖The name "cobalt" comes from the German word "kobalt" or "kobold," meaning "goblin." Miners gave it this name because they believed the metal, which was found alongside silver ores, was cursed, as it released toxic arsenic vapours during extraction.

❖Cobalt was discovered by Swedish chemist Georg Brandt in 1735 when he identified it as the element responsible for the blue colour in glass, distinguishing it from bismuth.

❖ Cobalt is famous for the vibrant blue colour it produces in glass and ceramics. Cobalt blue pigments have been used in art for centuries, including in ancient Egyptian jewellery and Chinese porcelain.

❖ Cobalt is one of only three naturally magnetic metals at room temperature (along with iron and nickel), making it useful in alloys for making strong magnets.

❖ Cobalt is a critical component in rechargeable lithium-ion batteries, which are used in smartphones, laptops, and electric vehicles. This has led to a high demand for cobalt in the tech industry.

❖ Cobalt is an important element in superalloys used for manufacturing turbine engines and aircraft parts, as it retains strength at high temperatures.

❖ Cobalt-60, a radioactive isotope of cobalt, is used in cancer treatment (radiotherapy) and for sterilising medical equipment. It is also used in industrial radiography for inspecting metal parts and welds.

❖ Cobalt is a vital part of vitamin B12, which is essential for red blood cell production & proper functioning of the nervous system.

❖ Much of the world's cobalt is mined in the Democratic Republic of Congo, which supplies over 60% of the global demand.

PHYSICAL PROPERTIES:

Physical Property	Description
Atomic Symbol	Co
Atomic Number	27
Atomic Mass	58.933194 amu
State at Room Temp	Solid (metal)
Color	Lustrous, silvery-blue
Odor	Odorless
Density	8.86 g/cm^3
Melting Point	1,495°C
Boiling Point	2,927°C
Solubility in Water	Insoluble
Conductivity	Good conductor of heat and electricity
Abundance in Earth	About 25 ppm in the Earth's crust

CHEMICAL PROPERTIES:

Chemical Property	Description
Reactivity	Cobalt is relatively stable in air, but oxidizes when heated, forming cobalt oxides
Oxidation States	+2, +3 (most common)
Electronegativity	1.88 (Pauling scale)
Reaction with Oxygen	Reacts with oxygen at high temperatures to form cobalt oxides (e.g., CoO, Co_3O_4)
Reaction with Acids	Dissolves in acids to form cobalt salts (e.g., cobalt chloride, $CoCl_2$)
Reaction with Water	Does not react with water at room temperature
Reaction with Halogens	Reacts with halogens to form cobalt halides (e.g., cobalt fluoride, CoF_2)
Flammability	Cobalt powder is flammable, but bulk cobalt is non-flammable
Toxicity	Cobalt compounds can be toxic in high concentrations, but it is essential for life in trace amounts
Biological Importance	Essential for vitamin B12 production, critical for red blood cell formation

NICKEL

Hi coin collectors! I'm Nickel, tough, shiny, and corrosion-resistant. From coins to batteries, I bring strength and durability to everyday objects.

INTERESTING FACTS ABOUT NICKEL

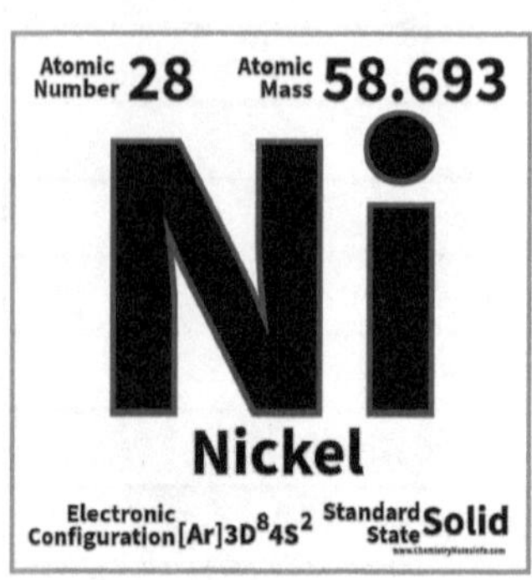

❖Nickel was discovered by Swedish chemist Axel Fredrik Cronstedt in 1751, who isolated it from the mineral niccolite, initially thinking it was a copper ore.

❖Nickel is best known for its use in stainless steel and other alloys, where it improves the strength, corrosion resistance, and toughness of metals.

❖Like iron & cobalt, nickel is ferromagnetic, meaning it can be magnetised. This property makes it useful in various electronic applications, including electromagnets and hard drives.

❖ Nickel has been used in coinage for over a century. The U.S. five-cent coin, called nickel, contains about 25% nickel and 75% copper.

❖ Nickel is a crucial component in several types of batteries, including nickel-cadmium (NiCd) and nickel-metal hydride (NiMH) batteries, commonly used in rechargeable devices.

❖ Nickel's resistance to oxidation & corrosion makes it ideal for use in harsh environments, such as in chemical plants & offshore oil rigs.

❖ Nickel is often used for electroplating metals to give them a shiny, corrosion-resistant surface. This is commonly seen in automotive parts, kitchen appliances, and coins.

❖ Nickel is found in significant amounts in iron meteorites. Its presence in meteorites is one of the clues that helped scientists learn more about the composition of Earth's core.

❖ Nickel is used as a catalyst in hydrogenation reactions, including in the production of margarine and shortening, where it helps in converting oils into solid fats.

❖ Nickel is believed to be one of the primary elements, along with iron, in Earth's core, contributing to the planet's magnetic field.

❖ Nickel has a relatively high melting point of 1,455°C.

"""

PHYSICAL PROPERTIES:

Physical Property	Description
Atomic Symbol	Ni
Atomic Number	28
Atomic Mass	58.6934 amu
State at Room Temp	Solid (metal)
Color	Lustrous, silvery-white
Odor	Odorless
Density	8.90 g/cm^3
Melting Point	1,455°C
Boiling Point	2,913°C
Solubility in Water	Insoluble
Conductivity	Good conductor of heat and electricity
Abundance in Earth	About 84 ppm in Earth's crust

CHEMICAL PROPERTIES:

Chemical Property	Description
Reactivity	Nickel is relatively unreactive; it forms a protective oxide layer in air
Oxidation States	+2 (most common), +3
Electronegativity	1.91 (Pauling scale)
Reaction with Oxygen	Forms nickel oxide (NiO) when heated in the presence of oxygen
Reaction with Acids	Dissolves in acids to form nickel salts (e.g., nickel chloride, $NiCl_2$)
Reaction with Water	Nickel does not react with water under normal conditions
Reaction with Halogens	Reacts with halogens to form nickel halides (e.g., nickel fluoride, NiF_2)
Flammability	Nickel powder is flammable, but bulk nickel is non-flammable
Toxicity	Nickel compounds can be toxic in high concentrations; long-term exposure can lead to respiratory issues
Biological Importance	Nickel is not essential for humans but is required by some enzymes in plants and bacteria

COPPER

Hello conductors! I'm Copper, the go-to element for electricity. Whether it's wiring your home or powering devices, I'm the metal that keeps energy flowing.

INTERESTING FACTS ABOUT COPPER

❖Copper was one of the first metals used by humans. Artifacts made from copper date back over 10,000 years, marking the beginning of the Copper Age.

❖The symbol for copper, Cu, comes from the Latin word cuprum, derived from Cyprium meaning "metal of Cyprus," since ancient Cyprus was a major source of copper.

❖Copper is highly conductive and is the primary material used in electrical wiring and electronics, making it essential in modern infrastructure.

❖ Copper is a key component in the alloy bronze (copper and tin), which was used extensively during the Bronze Age for tools, weapons, and statues.

❖ Copper has natural antibacterial properties. It can kill bacteria on surfaces, which is why it's often used in hospital door handles, bed rails, and other frequently touched objects.

❖ Copper is an essential trace mineral in the human body, playing a vital role in iron absorption, the formation of red blood cells, and maintaining the immune system.

❖ Copper is used in plumbing because it is resistant to corrosion, and it does not contaminate water. It has been a standard material in household plumbing for decades.

❖ The green colour of the Statue of Liberty comes from the copper it is made of. The copper has oxidised over time to form a layer of patina, which protects the statue from further corrosion.

❖ Historically, many countries used copper or copper alloys in their coinage. Today, U.S. pennies are made from zinc with a thin copper coating.

❖ Copper is 100% recyclable without any loss of quality. Most of the copper used today comes from recycled sources, contributing to sustainable practices.

PHYSICAL PROPERTIES:

Physical Property	Description
Atomic Symbol	Cu
Atomic Number	29
Atomic Mass	63.546 amu
State at Room Temp	Solid (metal)
Color	Reddish-brown, lustrous
Odor	Odorless
Density	8.96 g/cm^3
Melting Point	1,085°C
Boiling Point	2,562°C
Solubility in Water	Insoluble
Conductivity	Excellent conductor of heat and electricity
Abundance in Earth	About 60 ppm in the Earth's crust

CHEMICAL PROPERTIES:

Chemical Property	Description
Reactivity	Copper reacts slowly with oxygen in the air, forming a protective patina (copper oxide)
Oxidation States	+1, +2 (most common)
Electronegativity	1.90 (Pauling scale)
Reaction with Oxygen	Forms a layer of copper oxide (CuO or Cu_2O) when exposed to air
Reaction with Acids	Dissolves in strong acids like nitric acid, producing copper salts (e.g., $CuSO_4$)
Reaction with Water	Copper does not react with water, but reacts with carbon dioxide in moist air to form copper carbonate
Reaction with Halogens	Reacts with halogens to form copper halides (e.g., copper chloride, $CuCl_2$)
Flammability	Non-flammable in bulk form, but copper dust is flammable
Toxicity	Copper is non-toxic in small amounts, but high levels of copper exposure can lead to toxicity
Biological Importance	Essential trace element for humans, required for enzyme function and red blood cell formation
Isotopes	Two stable isotopes: Copper-63 (most abundant), Copper-65

ZINC

Hi protectors! I'm Zinc, keeping metals from rusting and boosting your immune system. From galvanising steel to treating colds, I'm always there to help.

INTERESTING FACTS ABOUT ZINC

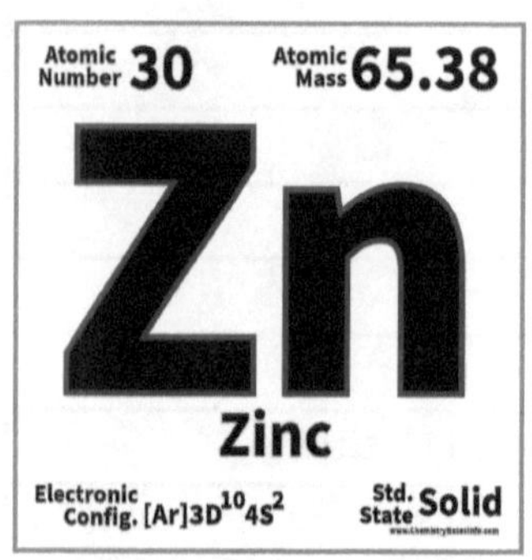

❖Zinc is an essential trace element for humans, playing a critical role in immune function, DNA synthesis, and wound healing. It's also necessary for the senses of taste and smell.

❖One of zinc's most important industrial applications is in galvanisation, where it is applied as a protective coating to prevent corrosion of iron and steel.

❖Zinc deficiency can lead to weakened immunity, delayed wound healing & impaired growth, especially in children. Often addressed through supplements and fortified foods.

❖ Zinc is a key component of brass, an alloy made with copper. Brass has been used for thousands of years for decorative items, musical instruments, and coins.

❖ Zinc is a moderately reactive metal that readily reacts with acids, producing hydrogen gas. This reaction is commonly used in laboratories to produce hydrogen.

❖ While zinc compounds have been used since ancient times, pure zinc was first isolated in India in the 13th century, and it was recognised as a distinct element by Andreas Marggraf in 1746.

❖ Zinc is a key component in several types of batteries, including zinc-carbon and zinc-air batteries, which are commonly used in flashlights, hearing aids, and remote controls.

❖ Zinc is a component of more than 300 enzymes in the human body, where it helps catalyse biochemical reactions, including the metabolism of nutrients.

❖ Zinc's resistance to corrosion, especially when used as a protective coating (galvanisation), is why it's commonly applied to metal structures exposed to the elements.

❖ Zinc is one of the most abundant metals in the Earth's crust, with major mining operations in countries like China, Australia & Peru.

PHYSICAL PROPERTIES:

Physical Property	Description
Atomic Symbol	Zn
Atomic Number	30
Atomic Mass	65.38 amu
State at Room Temp	Solid (metal)
Color	Bluish-silver, lustrous
Odor	Odorless
Density	7.14 g/cm^3
Melting Point	419.5°C
Boiling Point	907°C
Solubility in Water	Insoluble
Conductivity	Good conductor of heat and electricity
Abundance in Earth	About 75 ppm in Earth's crust

CHEMICAL PROPERTIES:

Chemical Property	Description
Reactivity	Zinc reacts with acids and strong alkalis, releasing hydrogen gas
Oxidation States	+2 (most common)
Electronegativity	1.65 (Pauling scale)
Reaction with Oxygen	Forms a thin, protective layer of zinc oxide (ZnO) when exposed to air
Reaction with Acids	Dissolves in acids like hydrochloric acid, producing zinc salts (e.g., $ZnCl_2$) and hydrogen gas
Reaction with Water	Zinc does not react with water under normal conditions
Reaction with Halogens	Reacts with halogens to form zinc halides (e.g., $ZnCl_2$)
Flammability	Zinc powder is flammable, but bulk zinc is non-flammable
Toxicity	Non-toxic in trace amounts but harmful if inhaled as a fine powder or ingested in large quantities
Biological Importance	Essential for humans; required for enzyme function, immune response, and cell growth
Isotopes	Five stable isotopes: Zinc-64 (most abundant), Zinc-66, Zinc-67, Zinc-68, Zinc-70

GALLIUM

Hello techies! I'm Gallium, known for melting in your hand but powering electronics. From LEDs to solar panels, I help brighten the world with tech!

INTERESTING FACTS ABOUT GALLIUM

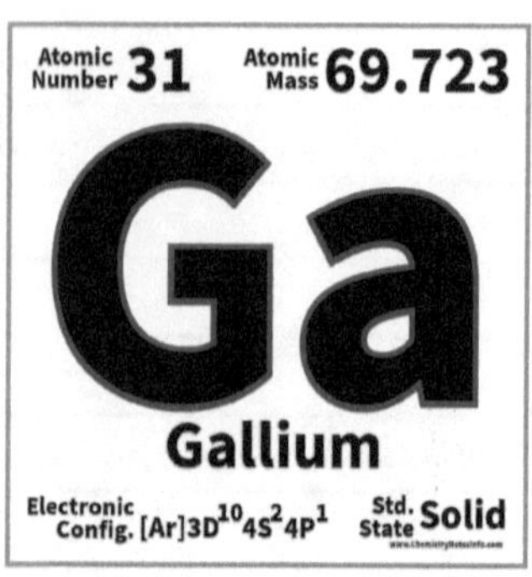

❖Gallium has a very low melting point of 29.76°C (85.57°F), which means it can melt from the warmth of your hand. This makes it unique among metals.

❖Gallium was discovered by French chemist Paul-Émile Lecoq de Boisbaudran in 1875 using spectroscopy. Its discovery confirmed predictions made by Dmitri Mendeleev about the existence of an element he called "eka-aluminum."

❖The name "gallium" comes from the Latin word Gallia, meaning France. It is also said to be a play on the discoverer's surname, Lecoq, which means "the rooster" in French, and gallus is the Latin word for rooster.

❖ Gallium is an important element in the electronics industry, particularly in the production of semiconductors, including gallium arsenide (GaAs), which is used in high-speed electronic devices and LEDs.

❖ Gallium is used in high-temperature thermometers because of its wide liquid range (it has a boiling point of 2,204°C). It is a safer alternative to mercury in certain applications.

❖ Unlike some other metals, gallium is non-toxic and can be safely handled. However, it can stain the skin and is difficult to clean up if spilled in liquid form.

❖ Gallium alloys are used to make mirrors for telescopes. These alloys can be deposited as a thin, reflective layer on glass or other materials.

❖ Unlike most substances, gallium expands when it freezes, which means it can crack containers if it's left to solidify in them.

❖ Gallium can wet glass and porcelain, sticking to their surfaces and making it tricky to handle in laboratories.

❖ Gallium in its liquid state has a higher viscosity than most other liquid metals, so it flows more slowly, almost like syrup.

PHYSICAL PROPERTIES:

Physical Property	Description
Atomic Symbol	Ga
Atomic Number	31
Atomic Mass	69.723 amu
State at Room Temp	Solid (but can melt at slightly warmer temperature)
Color	Silvery-blue, shiny when in solid form
Odor	Odorless
Density	5.91 g/cm³ (solid)
Melting Point	29.76°C
Boiling Point	2,204°C
Solubility in Water	Insoluble
Conductivity	Good conductor of electricity when in solid form, less so in liquid form
Abundance in Earth	About 19 ppm in Earth's crust

CHEMICAL PROPERTIES:

Chemical Property	Description
Reactivity	Gallium is relatively reactive; it forms a protective oxide layer when exposed to air
Oxidation States	+1, +3 (most common)
Electronegativity	1.81 (Pauling scale)
Reaction with Oxygen	Forms gallium oxide (Ga_2O_3) when exposed to oxygen in the air
Reaction with Acids	Reacts with strong acids, forming gallium salts and hydrogen gas
Reaction with Water	Gallium does not react with water under normal conditions
Reaction with Halogens	Reacts with halogens to form gallium halides (e.g., gallium chloride, $GaCl_3$)
Flammability	Gallium is non-flammable
Toxicity	Gallium is non-toxic in small amounts but can be harmful if ingested in large quantities
Biological Importance	Gallium has no known biological role, but gallium salts are used in some medical treatments
Isotopes	Two stable isotopes: Gallium-69 (most abundant), Gallium-71

GERMANIUM

Hi semiconductor fans! I'm Germanium, key to your fibre optics and transistors. I'm quietly powering the tech revolution, one signal at a time.

INTERESTING FACTS ABOUT GERMANIUM

❖Germanium was one of the elements predicted by Dmitri Mendeleev when he developed the periodic table. He called it "eka-silicon" before it was discovered because he knew it would share properties with silicon.

❖Germanium was discovered by German chemist Clemens Winkler in 1886. He found it while analyzing the mineral argyrodite, a silver sulfide ore.

❖Germanium is a metalloid and has semiconductor properties, making it crucial in the early development of transistors and other electronic components before silicon became more common.

❖ Germanium's ability to transmit infrared light makes it useful in infrared optics, including lenses, thermal imaging cameras, and night vision devices.

❖ Germanium is used in some types of light-emitting diodes (LEDs), which are essential for energy-efficient lighting and display technologies.

❖ Unlike other elements in its group, germanium is generally considered to be non-toxic, though some of its compounds can be harmful if ingested.

❖ Germanium has a high refractive index, which makes it useful in the creation of high-quality lenses and optical components.

❖ Germanium is relatively rare in the Earth's crust, with an abundance of about 1.5 ppm, but its applications in electronics, optics, and renewable energy make it highly valuable.

❖ Germanium is used in high-efficiency solar cells, particularly in space applications where maximum energy capture is critical, due to its ability to convert sunlight into electricity efficiently.

❖ Germanium is added to certain metal alloys to improve their corrosion resistance, especially in harsh environments.

PHYSICAL PROPERTIES:

Physical Property	Description
Atomic Symbol	Ge
Atomic Number	32
Atomic Mass	72.63 amu
State at Room Temp	Solid (metalloid)
Color	Grayish-white, metallic luster
Odor	Odorless
Density	5.323 g/cm^3
Melting Point	938.3°C
Boiling Point	2,820°C
Solubility in Water	Insoluble
Conductivity	Semiconductor, less conductive than metals but more so than insulators
Abundance in Earth	About 1.5 ppm in Earth's crust

CHEMICAL PROPERTIES:

Chemical Property	Description
Reactivity	Germanium is stable in air but slowly oxidizes at high temperatures
Oxidation States	+2, +4 (most common)
Electronegativity	2.01 (Pauling scale)
Reaction with Oxygen	Forms germanium dioxide (GeO_2) when heated in the presence of oxygen
Reaction with Acids	Reacts slowly with strong acids to form germanium salts
Reaction with Water	Germanium does not react with water under normal conditions
Reaction with Halogens	Reacts with halogens to form germanium halides (e.g., $GeCl_4$)
Flammability	Non-flammable in bulk form
Toxicity	Germanium is non-toxic in its elemental form, though some compounds can be toxic
Biological Importance	Germanium has no known biological role, though it has been studied for potential medicinal uses

ARSENIC

Hello chemists! I'm Arsenic, famous for my toxicity but also used in electronics and medicine. Handle me carefully—I'm full of surprises!

INTERESTING FACTS ABOUT ARSENIC

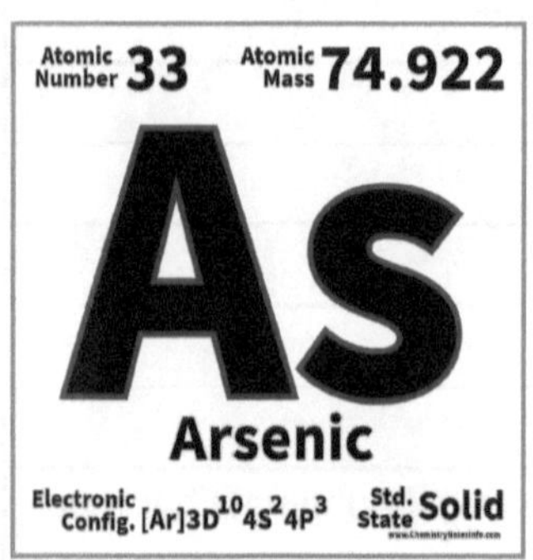

❖Arsenic is well-known for its toxicity. It has been historically used as a poison, earning a dark reputation in history as an agent of murder.

❖Arsenic has a long history, with its compounds known since ancient times. However, the element itself was first isolated in its pure form by the German alchemist Albertus Magnus around 1250.

❖Arsenic is used in the electronics industry, particularly in gallium arsenide (GaAs) semiconductors, which are critical components in high-speed devices and solar cells.

❖ Arsenic is naturally found in the Earth's crust in minerals like arsenopyrite and realgar. It can also be released into the environment through volcanic activity.

❖ Arsenic can exist in two main oxidation states: trivalent (As^{3+}) and pentavalent (As^{5+}). Trivalent arsenic compounds tend to be more toxic.

❖ In the 18th and 19th centuries, arsenic compounds were used in medicines to treat a variety of ailments, including syphilis and other infectious diseases.

❖ Arsenic was historically used in pesticides and herbicides, though its use has been restricted due to concerns about toxicity and environmental contamination.

❖ Arsenic compounds, such as copper arsenite (Scheele's green), were used in paints and wallpapers in the 19th century, though they were later banned due to their toxicity.

❖ Arsenic exposure, particularly through contaminated drinking water, has been linked to various forms of cancer, including skin, lung, and bladder cancer.

❖ While arsenic is toxic in high concentrations, trace amounts may play a role in metabolism for certain organisms, and some bacteria can use arsenic as part of their respiration process.

PHYSICAL PROPERTIES:

Physical Property	Description
Atomic Symbol	As
Atomic Number	33
Atomic Mass	74.9216 amu
State at Room Temp	Solid (metalloid)
Color	Metallic gray, yellow, or black forms
Odor	Odorless
Density	5.776 g/cm^3 (for gray arsenic)
Melting Point	Arsenic sublimes (turns from solid to gas) at 615°C (does not melt at normal pressure)
Boiling Point	613°C (sublimes)
Solubility in Water	Insoluble
Conductivity	Poor conductor of electricity
Abundance in Earth	About 1.8 ppm in Earth's crust

CHEMICAL PROPERTIES:

Chemical Property	Description
Reactivity	Arsenic is stable in air but reacts with oxygen when heated to form arsenic trioxide (As_2O_3)
Oxidation States	-3, +3, +5
Electronegativity	2.18 (Pauling scale)
Reaction with Oxygen	Forms arsenic trioxide (As_2O_3) or arsenic pentoxide (As_2O_5) when burned
Reaction with Acids	Reacts with strong oxidizing acids to form arsenic compounds
Reaction with Water	Does not react with water under normal conditions
Reaction with Halogens	Reacts with halogens to form arsenic halides (e.g., arsenic trichloride, $AsCl_3$)
Flammability	Non-flammable in bulk form, though arsenic dust is flammable
Toxicity	Highly toxic, both in elemental form and as arsenic compounds
Biological Importance	Toxic to most organisms, but trace amounts may have biological roles in some species
Isotopes	One stable isotope: Arsenic-75

SELENIUM

Hey photographers! I'm Selenium, essential for making solar cells and enhancing camera lenses. I'm all about capturing and converting light.

INTERESTING FACTS ABOUT SELENIUM

❖Selenium is an essential micronutrient for humans, playing a critical role in the metabolism, thyroid function & protecting the body from oxidative damage.

❖Selenium was discovered by Swedish chemist Jöns Jakob Berzelius in 1817. He found it while working with lead in a sulphuric acid plant and named it after the Greek word selene, meaning "moon."

❖Selenium is a semiconductor that exhibits photovoltaic properties, meaning it can convert light into electricity. This makes it useful in solar cells and light meters.

❖ Selenium is used to decolonize glass and to give it a red colour in some instances. It helps to counteract the greenish tint caused by iron impurities in the glass.

❖ Selenium's photoconductivity made it a key material in early xerographic (photocopying) processes, where light exposure alters its electrical conductivity.

❖ Selenium is essential for many animals, including humans, but it can be toxic in large amounts. It's particularly important in preventing oxidative stress in cells.

❖ Selenium deficiency can lead to a variety of health issues, including Keshan disease, a type of heart disease, and other immune-related problems.

❖ While selenium is necessary for health, excessive amounts can lead to selenium toxicity, also known as selenosis, which causes symptoms like gastrointestinal distress, hair loss, and neurological damage.

❖ Selenium is used in the production of rectifiers, which convert alternating current (AC) into direct current (DC), and in some types of semiconductors.

❖ Selenium poses antioxidant properties.

PHYSICAL PROPERTIES:

Physical Property	Description
Atomic Symbol	Se
Atomic Number	34
Atomic Mass	78.971 amu
State at Room Temp	Solid (non-metal)
Color	Comes in several allotropes, including gray metallic, red and black forms
Odor	Odorless
Density	4.81 g/cm³ (gray form)
Melting Point	221°C
Boiling Point	685°C
Solubility in Water	Insoluble
Conductivity	Photoconductive; conducts electricity when exposed to light
Abundance in Earth	About 0.05 ppm in Earth's crust

CHEMICAL PROPERTIES:

Chemical Property	Description
Reactivity	Selenium is reactive and forms compounds with metals and non-metals alike
Oxidation States	-2, +4, +6
Electronegativity	2.55 (Pauling scale)
Reaction with Oxygen	Forms selenium dioxide (SeO_2) when heated in air
Reaction with Acids	Reacts with strong acids, producing selenium salts
Reaction with Water	Selenium does not react with water under normal conditions
Reaction with Halogens	Reacts with halogens to form selenium halides (e.g., $SeCl_4$)
Flammability	Selenium is not flammable, though its compounds can be toxic
Toxicity	Selenium is toxic in large amounts, causing selenosis in humans and animals
Biological Importance	Essential for humans and animals in small amounts, critical for antioxidant enzyme function

BROMINE

Hi flame fighters! I'm Bromine, a liquid at room temperature and used in flame retardants. From safety gear to disinfectants, I've got your back!

INTERESTING FACTS ABOUT BROMINE

❖ Bromine is one of only two elements that are liquid at room temperature, other being mercury. It is reddish-brown, volatile liquid.

❖ Bromine was discovered by French chemist Antoine Jérôme Balard in 1826. He extracted it from sea salt water while studying salts from Mediterranean waters.

❖ The name "bromine" comes from the Greek word bromos, meaning "stench," due to its strong, unpleasant odour.

❖ Bromine is a highly reactive element and is part of the halogen group (Group 17). It readily forms compounds with many other elements, particularly metals.

❖ Bromine compounds, especially brominated flame retardants (BFRs), are widely used to reduce the flammability of textiles, plastics, and electronics.

❖ Bromine is sometimes used to disinfect swimming pools, hot tubs, and water supplies as an alternative to chlorine, particularly in warmer climates.

❖ Bromine is toxic and corrosive to human tissue. Its vapours can irritate eyes throat, and liquid bromine can cause severe skin burns.

❖ Bromine is naturally found in Earth's crust and seawater. Largest natural sources are evaporite deposits and salt lakes like Dead Sea.

❖ Silver bromide (AgBr) was historically important in photography. It is light-sensitive and was used in photographic plates and films.

❖ Bromine-containing compounds were historically used in sedatives, known as "bromides." Though no longer common, these sedatives were once used to treat anxiety and epilepsy.

❖ Methyl bromide was used as a pesticide and soil fumigant for decades. However, its use has been heavily restricted due to its harmful effects on the ozone layer.

PHYSICAL PROPERTIES:

Physical Property	Description
Atomic Symbol	Br
Atomic Number	35
Atomic Mass	79.904 amu
State at Room Temp	Liquid
Color	Reddish-brown (in liquid form); emits orange-brown vapor
Odor	Strong, choking odor
Density	3.1028 g/cm^3
Melting Point	$-7.2°C$
Boiling Point	$58.8°C$
Solubility in Water	Slightly soluble; forms bromine water (Br_2 in H_2O)
Conductivity	Non-conductive
Abundance in Earth	Found in trace amounts in Earth's crust; more abundant in seawater

CHEMICAL PROPERTIES:

Chemical Property	Description
Reactivity	Highly reactive; forms compounds readily with metals and non-metals
Oxidation States	$-1, +1, +3, +5, +7$
Electronegativity	2.96 (Pauling scale)
Reaction with Oxygen	Bromine does not react directly with oxygen but forms compounds with other elements in oxides
Reaction with Acids	Bromine reacts with concentrated acids, particularly with sulfuric acid
Reaction with Water	Slightly soluble in water, forming hydrobromic acid (HBr) and bromine water (Br_2 in H_2O)
Reaction with Halogens	Reacts with other halogens (chlorine, fluorine) to form interhalogen compounds
Flammability	Non-flammable, though it reacts violently with certain materials
Toxicity	Highly toxic; bromine vapors are harmful to respiratory system, and liquid bromine causes burns
Biological Importance	Plays no essential role in human biology, but small amounts are present in marine organisms

KRYPTON

Greetings, superheroes! I'm Krypton, glowing in high-intensity lamps and aiding lasers. While my name may be famous, I quietly help light up the world.

INTERESTING FACTS ABOUT KRYPTON

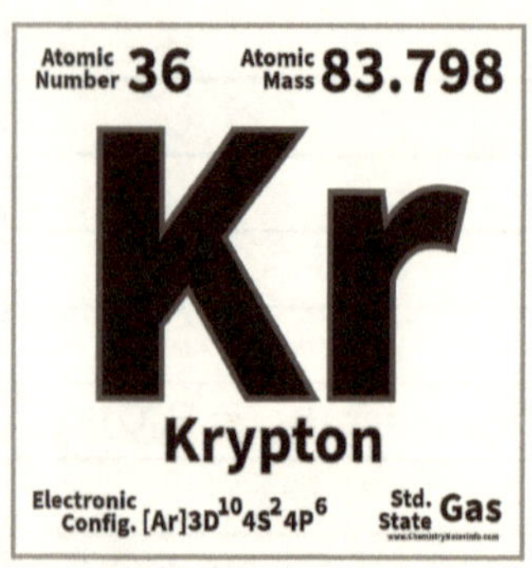

❖Krypton is a noble gas, meaning it is chemically inert and does not easily react with other elements. It is part of Group 18 of the periodic table.

❖Krypton was discovered by British chemists Sir William Ramsay and Morris Travers in 1898. It was found as a residual gas left behind after liquid air had almost entirely evaporated.

❖The name "krypton" comes from the Greek word kryptos, meaning "hidden," because it is difficult to isolate and detect due to its low concentration in the atmosphere.

❖ Krypton is a trace gas in Earth's atmosphere, making up about 1 part per million by volume.

❖ Krypton gas is used in certain types of lighting, including fluorescent lights and high-performance flashlights, where it helps to produce a bright, white light.

❖ Krypton is used in krypton fluoride (KrF) excimer lasers, which are used in precision applications such as eye surgery (LASIK) and semiconductor manufacturing.

❖ Although krypton is largely inert, it can form compounds with fluorine under extreme conditions, such as krypton difluoride (KrF_2).

❖ Krypton is used in some photographic flashes, particularly for high-speed photography because it produces bright, instantaneous light bursts.

❖ Krypton is about three times denser than air, which makes it a good insulator for windows and double-paned glass, helping to improve energy efficiency in buildings.

❖ Krypton gas is used in plasma displays to produce bright, vibrant colours, especially when mixed with other gases like neon and xenon.

PHYSICAL PROPERTIES:

Physical Property	Description
Atomic Symbol	Kr
Atomic Number	36
Atomic Mass	83.798 amu
State at Room Temp	Gas
Color	Colorless
Odor	Odorless
Density	3.749 g/L at 0°C and 1 atm
Melting Point	-157.37°C
Boiling Point	-153.415°C
Solubility in Water	Slightly soluble
Conductivity	Poor conductor of electricity and heat (as with most noble gases)
Abundance in Earth	Trace gas in the atmosphere, around 1 ppm

CHEMICAL PROPERTIES:

Chemical Property	Description
Reactivity	Krypton is chemically inert under most conditions but can form compounds with fluorine
Oxidation States	+2 (in rare compounds like KrF_2)
Electronegativity	3.00 (Pauling scale)
Reaction with Oxygen	Krypton does not react with oxygen under normal conditions
Reaction with Acids	Krypton is chemically inert and does not react with acids
Reaction with Water	Krypton does not react with water
Reaction with Halogens	Krypton can react with fluorine to form krypton difluoride (KrF_2) under extreme conditions
Flammability	Non-flammable
Toxicity	Non-toxic; inert gas
Biological Importance	No known biological role
Isotopes	Six stable isotopes, including Krypton-84 (most abundant)

RUBIDIUM

Hello timekeepers! I'm Rubidium, known for my use in atomic clocks. Precision is my game, and I keep things running smoothly and on time.

INTERESTING FACTS ABOUT RUBIDIUM

❖Rubidium is an alkali metal, meaning it belongs to Group 1 of the periodic table. It is soft and can be cut with a knife, like other alkali metals.

❖Rubidium was discovered by German chemists Robert Bunsen and Gustav Kirchhoff in 1861. They found it using a spectroscope, which showed bright red lines—hence its name from the Latin rubidus, meaning "deep red."

❖Rubidium is highly reactive, especially with water. When exposed to water, it reacts violently, producing hydrogen gas and heat, sometimes causing an explosion.

❖ Rubidium has a relatively low melting point (39.3°C or 102.7°F) for a metal, and it can melt if held in the palm of your hand.

❖ Rubidium is used in highly accurate atomic clocks. Though cesium-based atomic clocks are more common, rubidium clocks are still used in certain applications, including GPS systems.

❖ Although rubidium is not extremely rare, it is never found in pure form due to its high reactivity. It is often found mixed with other alkali metals in minerals like lepidolite and pollucite.

❖ When burned, rubidium produces a reddish-violet flame, making it useful in fireworks and flares for creating vibrant colours.

❖ Rubidium-82, a radioactive isotope of rubidium, is used in nuclear medicine for positron emission tomography (PET) imaging of the heart, helping diagnose coronary artery disease.

❖ Rubidium ions are being studied for potential use in ion propulsion engines for spacecraft. These engines use ionised particles to generate thrust in the vacuum of space.

❖ Rubidium compounds are sometimes used in specialty glass and ceramics manufacturing, improving their strength and durability.

PHYSICAL PROPERTIES:

Physical Property	Description
Atomic Symbol	Rb
Atomic Number	37
Atomic Mass	85.4678 amu
State at Room Temp	Solid (soft, silvery-white metal)
Color	Silvery-white
Odor	Odorless
Density	1.532 g/cm^3
Melting Point	39.31°C
Boiling Point	688°C
Solubility in Water	Reacts violently with water
Conductivity	Good conductor of electricity and heat, as with other alkali metals
Abundance in Earth	Around 90 ppm in Earth's crust

CHEMICAL PROPERTIES:

Chemical Property	Description
Reactivity	Highly reactive; reacts violently with water, acids, and non-metals like oxygen and halogens
Oxidation States	1
Electronegativity	0.82 (Pauling scale)
Reaction with Oxygen	Reacts rapidly with oxygen to form rubidium oxide (Rb_2O)
Reaction with Acids	Reacts with acids to form rubidium salts and release hydrogen gas
Reaction with Water	Reacts violently with water, producing rubidium hydroxide (RbOH) and hydrogen gas
Reaction with Halogens	Reacts with halogens to form rubidium halides (e.g., RbCl)
Flammability	Highly flammable; rubidium burns with a reddish-violet flame
Toxicity	Non-toxic in small amounts, but large doses can disrupt biological processes
Biological Importance	No known biological role, but can substitute for potassium in cellular functions

STRONTIUM

Hey fireworks fans! I'm Strontium, the element that gives those bright red bursts in the sky. I'm lighting up celebrations and medical imaging alike.

INTERESTING FACTS ABOUT STRONTIUM

❖Strontium is named after the village of Strontian in Scotland, where the mineral strontianite (strontium carbonate) was first discovered.

❖Strontium was discovered in 1790 by Scottish chemist Adair Crawford, but it wasn't isolated as a pure element until 1808 by Sir Humphry Davy using electrolysis.

❖Strontium salts, particularly strontium nitrate, are used in fireworks and flares to produce bright red flames, which are easily visible from a distance.

❖ Strontium has a chemical similarity to calcium, which causes it to be absorbed into bones. Strontium-90, a radioactive isotope, is particularly dangerous because it accumulates in bones and tissues.

❖ Strontium chloride is used in some types of toothpaste for people with sensitive teeth because it helps reduce tooth sensitivity by blocking the dental tubules.

❖ Strontium-90, a radioactive isotope, is a byproduct of nuclear reactions and is considered hazardous due to its long half-life and potential for causing bone cancer and leukaemia.

❖ This compound is used in some medications to treat osteoporosis. It helps strengthen bones by increasing bone density and reducing the risk of fractures.

❖ Strontium was once widely used in cathode ray tubes (CRTs) in older television sets to block X-ray emissions and enhance picture quality.

❖ Strontium is used in glass and ceramics manufacturing to produce colour effects and improve durability, especially in specialised optical lenses and glassware.

❖ Besides fireworks, strontium is used in pyrotechnics for creating bright, long-lasting flares, particularly for emergency signals and marine rescue operations.

PHYSICAL PROPERTIES:

Physical Property	Description
Atomic Symbol	Sr
Atomic Number	38
Atomic Mass	87.62 amu
State at Room Temp	Solid (soft, silvery-white metal)
Color	Silvery-white, but forms a yellow oxide layer when exposed to air
Odor	Odorless
Density	2.64 g/cm^3
Melting Point	777°C
Boiling Point	1,377°C
Solubility in Water	Reacts with water
Conductivity	Good conductor of heat and electricity
Abundance in Earth	About 370 ppm in Earth's crust, found mostly in strontianite and celestite minerals

CHEMICAL PROPERTIES:

Chemical Property	Description
Reactivity	Highly reactive; reacts with air, water, and acids
Oxidation States	+2 (forms Sr^{2+} ions)
Electronegativity	0.95 (Pauling scale)
Reaction with Oxygen	Reacts with oxygen to form strontium oxide (SrO), which gives a yellowish layer
Reaction with Acids	Reacts with acids to form strontium salts and release hydrogen gas
Reaction with Water	Reacts with water to form strontium hydroxide ($Sr(OH)_2$) and hydrogen gas
Reaction with Halogens	Reacts with halogens to form strontium halides (e.g., $SrCl_2$)
Flammability	Flammable, especially in powdered form
Toxicity	Non-toxic in stable form, but radioactive strontium-90 is highly dangerous
Biological Importance	No essential role in human biology, but used in treatments for bone-related diseases

YTTRIUM

Hello high-tech enthusiasts! I'm Yttrium, powering lasers, LEDs, and superconductors. I may be rare, but I help tech shine brighter and run smoother.

INTERESTING FACTS ABOUT YTTRIUM

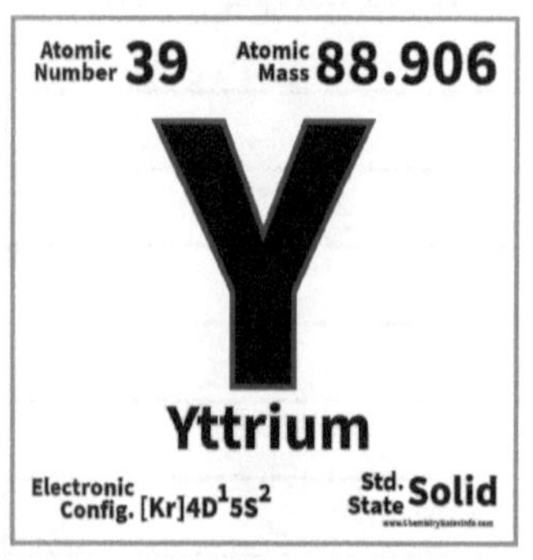

❖ Yttrium was discovered by Finnish chemist Johan Gadolin in 1794 in a mineral called gadolinite. It was named after the Swedish village of Ytterby, where the mineral was found.

❖ Yttrium is often classified as a rare earth element because it is found in the same ores as other rare earth elements, although it is not technically one.

❖ Yttrium compounds, such as yttrium oxide, are used in television screens and LEDs to create the red colour in phosphors for display technologies.

❖ Yttrium is a key component in certain high-temperature superconductors, specifically in yttrium barium copper oxide (YBCO), which has applications in MRI machines, power grids, and magnetic levitation trains.

❖ Yttrium aluminium garnet (YAG) lasers, often doped with neodymium, are used in surgery, dentistry, and industrial cutting and welding.

❖ Yttrium-90, a radioactive isotope, is used in cancer treatments, particularly for targeted radiation therapy, where it helps treat liver cancer by delivering radiation directly to tumours.

❖ Although yttrium is not a lanthanide, it is chemically similar to them and often occurs in nature with other lanthanides in mineral deposits.

❖ Yttrium is used in alloys with other metals, particularly in magnesium alloys, to improve corrosion resistance, making them useful in high-performance automotive and aerospace parts.

❖ Yttrium oxide is used in ceramic materials that are extremely resistant to heat and wear. These ceramics are used in applications like jet engine components & other high-temperature environments

❖ Yttrium is added to glass to improve its refractive index & strength.

PHYSICAL PROPERTIES:

Physical Property	Description
Atomic Symbol	Y
Atomic Number	39
Atomic Mass	88.90584 amu
State at Room Temp	Solid (silvery-white metal)
Color	Silvery-white
Odor	Odorless
Density	4.47 g/cm^3
Melting Point	1,522°C
Boiling Point	3,345°C
Solubility in Water	Insoluble
Conductivity	Good conductor of electricity and heat
Abundance in Earth	Around 33 ppm in Earth's crust

CHEMICAL PROPERTIES:

Chemical Property	Description
Reactivity	Moderately reactive; forms a protective oxide layer in air
Oxidation States	+3 (forms Y^{3+} ions)
Electronegativity	1.22 (Pauling scale)
Reaction with Oxygen	Reacts with oxygen to form yttrium oxide (Y_2O_3), which is used in various industrial applications
Reaction with Acids	Reacts with strong acids to form yttrium salts, such as yttrium chloride (YCl_3)
Reaction with Water	Slow reaction with water to form yttrium hydroxide ($Y(OH)_3$)
Reaction with Halogens	Reacts with halogens to form yttrium halides (e.g., YF_3, YCl_3)
Flammability	Non-flammable, but powdered yttrium can ignite at high temperatures
Toxicity	Non-toxic in solid form, but yttrium dust can be harmful if inhaled
Biological Importance	No known biological role; considered biologically inert
Isotopes	One stable isotope: Yttrium-89; Yttrium-90 is radioactive and used in medical applications

ZIRCONIUM

Hi space explorers! I'm Zirconium, super resistant to heat and corrosion. You'll find me in nuclear reactors and spacecraft, keeping things tough under pressure.

INTERESTING FACTS ABOUT ZIRCONIUM

❖Zirconium is most commonly known for its use in cubic zirconia, a synthetic gemstone that resembles diamonds. It is much cheaper than natural diamonds and widely used in jewellery.

❖The element zirconium is named after the mineral zircon, which is a zirconium silicate ($ZrSiO_4$) and has been used as a gemstone for centuries.

❖Zirconium was discovered by German chemist Martin Heinrich Klaproth in 1789. He identified it in the mineral zircon, though it wasn't isolated as a pure element until 1824 by Swedish chemist Jöns Jakob Berzelius.

❖ Zirconium is highly resistant to corrosion, especially in acidic and basic environments. This makes it ideal for use in chemical processing equipment and nuclear reactors.

❖ Zirconium is used in the nuclear industry to make cladding for fuel rods. Its low neutron absorption cross-section allows it to be used safely in nuclear reactors without interfering with the nuclear reaction.

❖ Because zirconium is non-toxic and resists corrosion, it is used in medical applications, such as dental implants and prosthetic devices.

❖ Zirconium is often alloyed with other metals like niobium and titanium to create heat-resistant materials for aerospace applications, including jet engines and spacecraft parts.

❖ Zirconium has been detected in lunar rocks brought back from the Moon by the Apollo missions, giving clues to the Moon's formation and geologic history.

❖ Zirconium compounds burn with a bright white flame, which is why they are used in fireworks, flashbulbs, and military flares.

❖ Zircon crystals are used in radiometric dating, particularly uranium-lead dating, to determine the age of ancient rocks and the Earth itself.

PHYSICAL PROPERTIES:

Physical Property	Description
Atomic Symbol	Zr
Atomic Number	40
Atomic Mass	91.224 amu
State at Room Temp	Solid (hard, silver-grey metal)
Color	Silvery-white
Odor	Odourless
Density	6.52 g/cm³
Melting Point	1,855°C
Boiling Point	4,409°C
Solubility in Water	Insoluble
Conductivity	Good conductor of heat and electricity
Abundance in Earth	Around 165 ppm in Earth's crust, found primarily in zircon and baddeleyite minerals

CHEMICAL PROPERTIES:

Chemical Property	Description
Reactivity	Reacts with oxygen, nitrogen, and halogens at high temperatures
Oxidation States	+4 is the most common oxidation state
Electronegativity	1.33 (Pauling scale)
Reaction with Oxygen	Forms a protective layer of zirconium oxide (ZrO_2) when exposed to air
Reaction with Acids	Resists attack by most acids but can dissolve in hydrofluoric acid
Reaction with Water	Reacts with steam at high temperatures to form zirconium dioxide (ZrO_2) and hydrogen gas
Reaction with Halogens	Reacts with halogens at high temperatures to form zirconium halides (e.g., $ZrCl_4$)
Flammability	Non-flammable in bulk, but powdered zirconium is pyrophoric
Toxicity	Non-toxic and biocompatible, making it useful in medical implants
Biological Importance	No known biological role
Isotopes	Five stable isotopes, the most common being Zirconium-90

NIOBIUM

Hello builders! I'm Niobium, a key player in making super strong steel for bridges, pipelines, and even superconductors. Strength and flexibility, that's my style.

INTERESTING FACTS ABOUT NIOBIUM

❖Niobium was originally called "columbium" when it was first discovered in 1801 by Charles Hatchett, after mineral columbite, but was later renamed "niobium" in 1949.

❖The name "niobium" comes from Niobe, the daughter of Tantalus in Greek mythology. The element tantalum, which is chemically similar to niobium, was named after Tantalus.

❖Niobium is a soft, malleable, and ductile metal that can be easily drawn into wires or shaped into thin sheets, making it useful in various industrial applications.

❖ Niobium is a key component in superconducting magnets, especially niobium-titanium alloys, which are used in MRI machines and particle accelerators.

❖ Niobium is added to steel to increase its strength, toughness, and resistance to heat. Niobium-containing steel is used in pipelines, automobiles, and construction.

❖ Niobium alloys are used in the aerospace industry for jet engines and rockets because they can withstand high temperatures and resist oxidation.

❖ Niobium is highly resistant to corrosion, especially by acids, making it useful for chemical processing equipment and laboratory instruments.

❖ Niobium is used in alloys for nuclear reactors due to its ability to absorb very few neutrons, making it ideal for maintaining the integrity of reactor components.

❖ Like titanium, niobium is biocompatible and used in medical implants and prosthetics because it doesn't react with bodily fluids or tissues.

❖ Niobium capacitors are used in electronic devices as alternative to tantalum capacitors due to similar properties but lower cost.

PHYSICAL PROPERTIES:

Physical Property	Description
Atomic Symbol	Nb
Atomic Number	41
Atomic Mass	92.90638 amu
State at Room Temp	Solid (lustrous, silvery-gray metal)
Color	Silvery-gray
Odor	Odorless
Density	8.57 g/cm^3
Melting Point	2,468°C
Boiling Point	4,927°C
Solubility in Water	Insoluble
Conductivity	Good conductor of electricity and heat
Abundance in Earth	Approximately 20 ppm in Earth's crust, found mainly in minerals like columbite and pyrochlore

CHEMICAL PROPERTIES:

Chemical Property	Description
Reactivity	Stable at room temperature, but reacts with oxygen, nitrogen, and carbon at high temperatures
Oxidation States	+5 is the most common oxidation state, although it can also exist in +3 and +4 states
Electronegativity	1.6 (Pauling scale)
Reaction with Oxygen	Reacts with oxygen at high temperatures to form niobium pentoxide (Nb_2O_5)
Reaction with Acids	Resistant to many acids, but dissolves in hydrofluoric acid
Reaction with Water	Stable in water at normal temperatures
Reaction with Halogens	Reacts with halogens to form niobium halides (e.g., $NbCl_5$, NbF_5)
Flammability	Non-flammable in bulk, but fine niobium powder can be flammable
Toxicity	Considered non-toxic in its pure metallic form
Biological Importance	No known biological role

MOLYBDENUM

Hi mechanics! I'm Molybdenum, a critical component in making strong, heat-resistant steel. From engines to electronics, I keep things running under stress.

INTERESTING FACTS ABOUT MOLYBDENUM

❖Molybdenum is an essential trace element for humans, plants, and animals. It is part of important enzymes that facilitate nitrogen fixation and the metabolism of sulphur-containing amino acids.

❖Molybdenum was discovered by Swedish chemist Carl Wilhelm Scheele in 1778, though it was confused with graphite and lead ores for a long time. It was first isolated as a metal by Peter Jacob Hjelm in 1781.

❖The name "molybdenum" comes from the Greek word "molybdos," meaning lead, as it was initially mistaken for a lead ore.

❖ Molybdenum is added to steel to increase its strength, toughness, and heat resistance. These molybdenum steel alloys are used in construction, shipbuilding, and oil pipelines.

❖ Molybdenum has one of the highest melting points of all elements (2,623°C), making it ideal for use in high-temperature environments like furnaces and jet engines.

❖ Molybdenum compounds are used as catalysts in the oil industry, particularly in petroleum refining, where they help remove sulphur from fuels.

❖ It is used in thin-film transistors and other electronics due to its good conductivity and resistance to oxidation, particularly in displays like LCDs.

❖ Molybdenum is a crucial component of nitrogenase, an enzyme that allows nitrogen-fixing bacteria to convert atmospheric nitrogen into a usable form for plants.

❖ Molybdenum resists corrosion and oxidation at high temperatures, which makes it useful in harsh chemical environments, including nuclear power plants.

❖ Molybdenum is used in radiotherapy, especially Molybdenum-99, which decays to produce Technetium-99m.

PHYSICAL PROPERTIES:

Physical Property	Description
Atomic Symbol	Mo
Atomic Number	42
Atomic Mass	95.95 amu
State at Room Temp	Solid (silvery-gray metal)
Color	Silvery-white
Odor	Odorless
Density	10.28 g/cm^3
Melting Point	2,623°C
Boiling Point	4,639°C
Solubility in Water	Insoluble
Conductivity	Good conductor of electricity and heat
Abundance in Earth	Approximately 1.2 ppm in Earth's crust, often found in molybdenite ores

CHEMICAL PROPERTIES:

Chemical Property	Description
Reactivity	Reacts with oxygen, sulfur, and halogens at high temperatures
Oxidation States	+6 is the most common oxidation state, but can also exist in +4 and +2 states
Electronegativity	2.16 (Pauling scale)
Reaction with Oxygen	Forms molybdenum trioxide (MoO_3) when heated in air
Reaction with Acids	Molybdenum is resistant to most acids at room temperature but reacts with hot acids
Reaction with Water	Does not react with water under normal conditions
Reaction with Halogens	Reacts with halogens to form molybdenum halides (e.g., $MoCl_5$, MoF_6)
Flammability	Non-flammable
Toxicity	Low toxicity; essential in small amounts, but excessive exposure can be harmful
Biological Importance	Essential trace element in enzymes involved in nitrogen fixation and sulfur metabolism
Isotopes	Several isotopes, with Molybdenum-98 being the most abundant stable isotope

TECHNETIUM

Greetings radiologists! I'm Technetium, the first man-made element, and I'm invaluable in medical imaging. I help doctors see inside the human body like never before!

INTERESTING FACTS ABOUT TECHNETIUM

❖Technetium was the first element to be artificially produced in a laboratory. It was created in 1937 by Italian scientists Carlo Perrier and Emilio Segrè, who extracted it from the molybdenum that had been bombarded with deuterons.

❖All isotopes of technetium are radioactive. It is one of the few elements with no stable isotopes, making it unique among the first 83 elements on the periodic table.

❖Technetium-99m, a metastable isotope, is widely used in medical imaging, particularly in nuclear medicine for diagnostic scans. It is used in millions of procedures each year for imaging organs such as the heart, brain, and bones.

❖ Technetium-99m has a half-life of only six hours, making it ideal for medical applications, as it decays quickly and minimises radiation exposure to patients.

❖ Despite its artificial creation on Earth, technetium has been detected in the atmospheres of certain stars, which suggests that stellar nucleosynthesis is responsible for producing it.

❖ Technetium can be added in small amounts to steel to prevent corrosion, particularly in environments where radiation is present, such as nuclear reactors.

❖ Technetium-99, a byproduct of nuclear fission in reactors, has a long half-life of about 211,000 years and is a major concern in nuclear waste management.

❖ Technetium was predicted by Dmitri Mendeleev, who called it "eka-manganese" before its actual discovery, as there was a gap in the periodic table where technetium now resides.

❖ Technetium is extremely rare in nature. It is found only in minute quantities in uranium ores due to spontaneous fission, making it largely a synthetic element.

PHYSICAL PROPERTIES:

Physical Property	Description
Atomic Symbol	Tc
Atomic Number	43
Atomic Mass	98 amu (most stable isotope, Tc-99)
State at Room Temp	Solid (silvery-gray metal)
Color	Metallic gray
Odor	Odorless
Density	11 g/cm^3
Melting Point	2,157°C
Boiling Point	4,265°C
Solubility in Water	Insoluble
Conductivity	Good conductor of electricity and heat
Abundance in Earth	Extremely rare, mostly synthetic, found only in trace amounts in uranium ores

CHEMICAL PROPERTIES:

Chemical Property	Description
Reactivity	Reacts with oxygen, sulfur, and halogens, especially at high temperatures
Oxidation States	Most common oxidation states are +7, +4, and +5
Electronegativity	1.9 (Pauling scale)
Reaction with Oxygen	Reacts with oxygen at elevated temperatures to form technetium oxides (e.g., Tc_2O_7)
Reaction with Acids	Dissolves in concentrated nitric acid & sulphuric acid
Reaction with Water	Stable in water under normal conditions
Reaction with Halogens	Reacts with halogens to form technetium halides (e.g., $TcCl_4$)
Flammability	Non-flammable
Toxicity	Radioactive, with significant health risks if ingested or inhaled in sufficient quantities
Biological Importance	No known biological role
Isotopes	Over 30 isotopes, with Technetium-99 and Technetium-99m being the most important in applications

RUTHENIUM

Hello catalysts! I'm Ruthenium, a rare metal making waves in electronics and catalysis. From hard disks to solar cells, I help tech perform better.

INTERESTING FACTS ABOUT RUTHENIUM

❖Ruthenium is part of the platinum group metals (PGMs), a set of six precious metals that include platinum, palladium, rhodium, iridium, and osmium. These elements are known for their catalytic properties and resistance to corrosion.

❖Ruthenium was discovered in 1844 by Russian chemist Karl Ernst Claus, who named it after his homeland, "Ruthenia," the Latin name for Russia.

❖Ruthenium is one of the rarest elements on Earth, found in very low concentrations in platinum ores and nickel mining by-products.

❖ Ruthenium is used in the electronics industry, especially in the production of thick-film resistors and electrical contacts, due to its excellent conductive properties.

❖ Ruthenium is widely used as a catalyst in chemical reactions, particularly in the hydrogenation of organic compounds and in the ammonia synthesis for fertilisers (Haber-Bosch process).

❖ When alloyed with platinum and palladium, ruthenium increases their hardness and corrosion resistance, making it valuable in jewellery and high-precision instruments.

❖ Some ruthenium compounds are being studied for their potential use in cancer treatment, particularly as alternatives to platinum-based chemotherapy drugs like cisplatin.

❖ Ruthenium is highly resistant to corrosion & oxidation, even at high temperatures, which makes it useful in industrial applications.

❖ When alloyed with titanium, ruthenium makes the metal more resistant to corrosion, especially in harsh chemical environments, such as in seawater and in chemical plants.

❖ Despite its strong catalytic activity, ruthenium is relatively chemically inactive in many environments, giving it a long lifespan in various industrial applications.

PHYSICAL PROPERTIES:

Physical Property	Description
Atomic Symbol	Ru
Atomic Number	44
Atomic Mass	101.07 amu
State at Room Temp	Solid (hard, silvery-white metal)
Color	Silvery-white
Odor	Odorless
Density	12.45 g/cm^3
Melting Point	2,334°C
Boiling Point	4,150°C
Solubility in Water	Insoluble
Conductivity	Good conductor of electricity and heat
Abundance in Earth	Extremely rare, found in trace amounts in platinum ores and nickel mining by-products

CHEMICAL PROPERTIES:

Chemical Property	Description
Reactivity	Stable in air and water at room temperature but reacts with halogens and acids at high temperatures
Oxidation States	Common oxidation states are +3, +4, +6, and +8
Electronegativity	2.2 (Pauling scale)
Reaction with Oxygen	Forms ruthenium dioxide (RuO_2) when heated in air
Reaction with Acids	Resistant to most acids, but dissolves in aqua regia
Reaction with Water	Stable in water under normal conditions
Reaction with Halogens	Reacts with halogens to form ruthenium halides (e.g., $RuCl_3$, RuF_5)
Flammability	Non-flammable
Toxicity	Ruthenium compounds are toxic and should be handled with care
Biological Importance	No known biological role, but some ruthenium compounds are being researched for medical applications
Isotopes	Several isotopes, with Ruthenium-101 and Ruthenium-102 being the most stable

RHODIUM

Hi shine seekers! I'm Rhodium, one of the rarest and most reflective metals. You'll find me in jewellery and catalytic converters, bringing brilliance and efficiency.

INTERESTING FACTS ABOUT RHODIUM

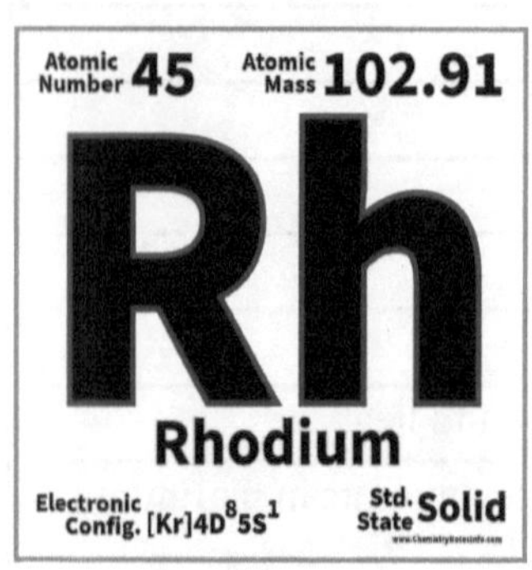

❖Rhodium is one of the rarest and most expensive metals in the world. It is significantly rarer than gold and platinum and is highly valuable due to its scarcity and desirable properties.

❖Rhodium was discovered by English chemist William Hyde Wollaston in 1803, shortly after his discovery of palladium. He named it "rhodium" due to rose colour of its salts (from Greek "rhodon," meaning rose).

❖Rhodium is primarily used in catalytic converters in automobiles to reduce harmful emissions like nitrogen oxides (NOx) from exhaust gases. This use accounts for the majority of global rhodium demand.

❖ Rhodium is one of the most reflective metals, which makes it ideal for plating surfaces to create highly reflective finishes. It is often used in mirrors, searchlights, and jewellery for this purpose.

❖ Rhodium is frequently used as a plating for white gold and silver jewellery to give them a brighter, more durable finish that is resistant to tarnishing and scratches.

❖ Rhodium is chemically inert and does not easily react with most substances. This makes it an excellent protective coating for metals that are exposed to corrosive environments.

❖ Rhodium has a high melting point of 1,964°C, making it suitable for high-temperature applications like furnace components and thermocouples.

❖ Rhodium is rarer than gold and platinum, with an estimated global production of only about 30 tons per year, making it one of the most sought-after precious metals.

❖ The Nobel Prize medals awarded from 1931 onward have been made from gold, but they are plated with a thin layer of rhodium for protection and durability.

PHYSICAL PROPERTIES:

Physical Property	Description
Atomic Symbol	Rh
Atomic Number	45
Atomic Mass	102.91 amu
State at Room Temp	Solid (hard, silvery-white metal)
Color	Silvery-white
Odor	Odorless
Density	12.41 g/cm^3
Melting Point	1,964°C
Boiling Point	3,695°C
Solubility in Water	Insoluble
Conductivity	Excellent conductor of electricity and heat
Abundance in Earth	Extremely rare, mostly found in platinum and nickel ores

CHEMICAL PROPERTIES:

Chemical Property	Description
Reactivity	Chemically inert under most conditions
Oxidation States	Common oxidation states are +3 and +1
Electronegativity	2.28 (Pauling scale)
Reaction with Oxygen	Does not react with oxygen at room temperature but forms oxides at higher temperatures
Reaction with Acids	Resistant to most acids, but dissolves in aqua regia (a mixture of nitric and hydrochloric acids)
Reaction with Water	Stable in water
Reaction with Halogens	Reacts with halogens at high temperatures to form rhodium halides
Flammability	Non-flammable
Toxicity	Rhodium compounds can be toxic if inhaled or ingested in large amounts, but elemental rhodium is inert
Biological Importance	No known biological role
Isotopes	Has one naturally occurring isotope: Rhodium-103

PALLADIUM

Hey auto enthusiasts! I'm Palladium, a key element in catalytic converters, helping reduce vehicle emissions. I'm all about keeping the air cleaner.

INTERESTING FACTS ABOUT PALLADIUM

❖Palladium is primarily used in catalytic converters in automobiles, where it helps reduce harmful emissions by converting toxic gases like carbon monoxide and nitrogen oxides into less harmful substances.

❖Palladium was discovered in 1803 by English chemist William Hyde Wollaston. He named it after the asteroid Pallas, which had been discovered two years earlier.

❖Palladium has a unique ability to absorb hydrogen gas, with the capacity to absorb up to 900 times its own volume. This makes it useful in hydrogen storage and purification technologies.

❖ Palladium is sometimes used in coins and bullion as an alternative to gold, silver, and platinum. It is considered a precious metal and is valued for its scarcity and industrial uses.

❖ Palladium is used in jewellery, especially as an alloy with gold to create white gold. It gives a bright finish without the need for rhodium plating.

❖ Palladium is used in fuel cells, where hydrogen is converted into electricity in a clean and efficient way. This technology is considered key for future energy systems.

❖ Palladium is an important catalyst in chemical reactions, especially in organic chemistry for processes like hydrogenation and cross-coupling reactions (e.g., the Suzuki and Heck reactions).

❖ Palladium is highly resistant to corrosion and tarnishing, making it a valuable metal for industrial applications and jewellery.

❖ The 2010 Nobel Prize in Chemistry was awarded for work involving palladium-catalysed reactions, highlighting its significance in modern organic synthesis.

❖ It is used in dentistry for crowns & bridges, as well as in certain medical devices due to its biocompatibility & corrosion resistance.

PHYSICAL PROPERTIES:

Physical Property	Description
Atomic Symbol	Pd
Atomic Number	46
Atomic Mass	106.42 amu
State at Room Temp	Solid (soft, silvery-white metal)
Color	Silvery-white
Odor	Odorless
Density	12.02 g/cm³
Melting Point	1,554.9°C
Boiling Point	2,963°C
Solubility in Water	Insoluble
Conductivity	Excellent conductor of electricity and heat
Abundance in Earth	Rare, found in platinum group ores and nickel deposits

CHEMICAL PROPERTIES:

Chemical Property	Description
Reactivity	Chemically stable, resistant to tarnishing and corrosion in air
Oxidation States	Common oxidation states are +2 and +4
Electronegativity	2.20 (Pauling scale)
Reaction with Oxygen	Does not react with oxygen at room temperature but forms palladium oxide (PdO) when heated
Reaction with Acids	Dissolves in concentrated nitric acid and sulfuric acid
Reaction with Water	Stable in water
Reaction with Halogens	Reacts with halogens at high temperatures to form palladium halides
Flammability	Non-flammable
Toxicity	Elemental palladium is non-toxic, but palladium compounds can be harmful in large amounts
Biological Importance	No known biological role, but used in dental and medical implants due to its biocompatibility
Isotopes	Palladium has six stable isotopes, with Palladium-106 being the most abundant

SILVER

Hello treasure hunters! I'm Silver, prized for my beauty and conductivity. From jewellery to electronics, I'm the shiny metal that's both glamorous and functional.

INTERESTING FACTS ABOUT SILVER

❖Silver has been used by humans for thousands of years, dating back to ancient civilisations. It was highly valued for its beauty and utility, often used for currency, jewellery, and tableware.

❖The symbol for silver, Ag, comes from the Latin word "argentum," which means silver and is derived from the Proto-Indo-European root "arg-" meaning "to shine."

❖Silver is the best conductor of electricity and heat among all metals. Its high conductivity makes it valuable in electrical applications and components.

❖ Silver has natural antimicrobial properties and is used in medical applications, such as wound dressings, coatings for medical devices, and water purification systems.

❖ When exposed to air and moisture, silver reacts with sulphur compounds, leading to tarnishing. This can be removed with special cleaning solutions or polishes.

❖ Silver halides were essential in traditional photography, forming light-sensitive compounds used in photographic film and paper. Although digital photography has largely replaced this, silver is still used in some specialised applications.

❖ Silver is widely used in jewelry and high-quality tableware. Sterling silver, an alloy made of 92.5% silver and 7.5% other metals (usually copper), is a common standard for jewellery and cutlery.

❖ The production of silver often involves mining and refining processes. Major silver-producing countries include Mexico, Peru, China, and Australia.

❖ Silver is considered a precious metal and is often used as an investment. Investors purchase silver bullion coins, bars, and exchange-traded funds (ETFs) as a hedge against inflation.

❖ Silver is used in various space applications, including solar panels.

PHYSICAL PROPERTIES:

Physical Property	Description
Atomic Symbol	Ag
Atomic Number	47
Atomic Mass	107.87 amu
State at Room Temp	Solid (shiny, metallic)
Color	Lustrous white
Odor	Odorless
Density	10.49 g/cm^3
Melting Point	961.8°C
Boiling Point	2,162°C
Solubility in Water	Insoluble
Conductivity	Excellent conductor of electricity and heat
Abundance in Earth	Relatively rare, found in ores such as argentite and in native form

CHEMICAL PROPERTIES:

Chemical Property	Description
Reactivity	Relatively unreactive, but can tarnish in the presence of sulfur compounds
Oxidation States	Common oxidation states are +1 and +2
Electronegativity	1.93 (Pauling scale)
Reaction with Oxygen	Forms silver oxide (Ag_2O) when heated in the presence of oxygen
Reaction with Acids	Dissolves in nitric acid and concentrated sulfuric acid
Reaction with Water	Stable in water
Reaction with Halogens	Reacts with halogens to form silver halides (e.g., AgCl, AgBr)
Flammability	Non-flammable
Toxicity	Elemental silver is generally non-toxic, but silver compounds can be harmful in large amounts
Biological Importance	Silver ions are toxic to bacteria, making silver useful in medical applications
Isotopes	Silver has two stable isotopes: Silver-107 and Silver-109

CADMIUM

Hi protectors! I'm Cadmium, used in batteries and solar panels. While toxic, I'm invaluable in keeping your gadgets powered up and protecting metals from corrosion.

INTERESTING FACTS ABOUT CADMIUM

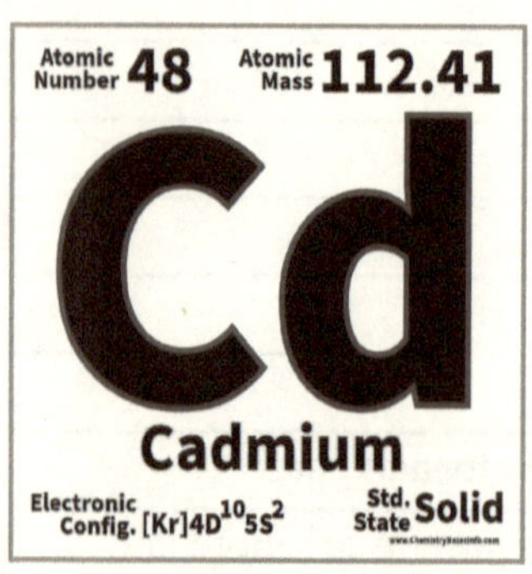

❖Cadmium is primarily used in industrial applications, including the manufacturing of batteries, pigments, coatings, and stabilizers for plastics.

❖Cadmium was discovered in 1817 by Swedish chemist Friedrich Strohmeyer while he was studying zinc ores. It was named after the Latin word "cadmia," which refers to a type of zinc ore.

❖Cadmium is commonly used in nickel-cadmium (NiCd) rechargeable batteries, which are known for their durability and ability to withstand deep discharges. However, their use has declined due to environmental concerns.

❖ Cadmium compounds are used as pigments in paints, plastics, and ceramics, providing vibrant colors like yellow, red, and orange. Cadmium sulfide (CdS) is a popular yellow pigment.

❖ Cadmium is toxic and poses health risks to humans, especially when inhaled or ingested. It can cause kidney damage, bone disease, and other health issues, which has led to strict regulations regarding its use.

❖ Cadmium can accumulate in the environment and bioaccumulate in living organisms, particularly in aquatic life. This has raised concerns about its impact on ecosystems and human health.

❖ Cadmium is used to create certain alloys, particularly in applications requiring low friction and corrosion resistance, such as in aerospace and electronics.

❖ Cadmium is used in electroplating to provide a protective coating for steel and iron products, enhancing their resistance to corrosion.

❖ Cadmium has applications in nuclear reactors due to its ability to absorb neutrons, helping to control the nuclear fission process.

❖ While cadmium is not considered an essential nutrient, trace amounts can be found in some organisms.

PHYSICAL PROPERTIES:

Physical Property	Description
Atomic Symbol	Cd
Atomic Number	48
Atomic Mass	112.41 amu
State at Room Temp	Solid (soft, malleable, silvery-white metal)
Color	Silvery-blue
Odor	Odorless
Density	8.65 g/cm^3
Melting Point	321.1°C
Boiling Point	767°C
Solubility in Water	Insoluble
Conductivity	Fair conductor of electricity and heat
Abundance in Earth	Relatively rare, found mainly in zinc ores and as a by-product of zinc refining

CHEMICAL PROPERTIES:

Chemical Property	Description
Reactivity	Reactive metal, especially in the presence of acids
Oxidation States	Common oxidation states are +1 and +2
Electronegativity	1.69 (Pauling scale)
Reaction with Oxygen	Forms cadmium oxide (CdO) when heated in air
Reaction with Acids	Dissolves in dilute acids, producing hydrogen gas and cadmium salts
Reaction with Water	Does not react with water under normal conditions
Reaction with Halogens	Reacts with halogens to form cadmium halides (e.g., $CdCl_2$)
Flammability	Non-flammable but can produce toxic fumes upon combustion
Toxicity	Cadmium compounds are toxic and can cause health problems upon exposure
Biological Importance	Not considered essential for any biological functions; generally harmful
Isotopes	Cadmium has several isotopes, with Cadmium-114 & Cadmium-116 being most stable

INDIUM

Hello touchscreens! I'm Indium, helping to make your devices touch-responsive and transparent. From smartphones to flat screens, I'm at the forefront of modern tech.

INTERESTING FACTS ABOUT INDIUM

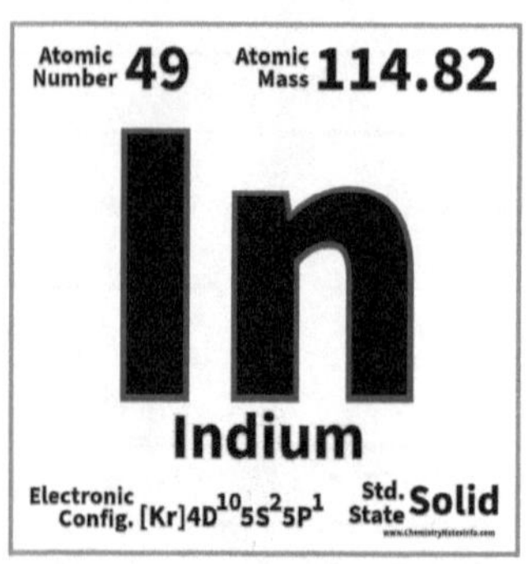

❖Indium was discovered in 1863 by German chemist Ferdinand Reich and his student Hieronymous Theodor Richter. They identified it while analysing zinc ores and named it after the indigo colour of its spectral lines.

❖Indium is a soft, malleable metal that can be easily shaped and drawn into wires. It is so soft that it can be scratched with a fingernail.

❖Indium has a relatively low melting point of 156.6°C, making it useful in low-temperature applications.

❖ Indium is used in electronics, particularly in the production of indium tin oxide (ITO), a transparent conductive oxide widely used in touchscreens, flat-panel displays, and solar cells.

❖ Indium is used as an alloying agent to improve the properties of other metals, such as lead, to make them more resistant to corrosion and to lower the melting point.

❖ Unlike some other metals, indium is considered relatively nontoxic, which makes it a safer choice for various applications, including electronics.

❖ Indium has a high coefficient of thermal expansion, which means it expands significantly when heated. This property makes it useful in applications that require a tight seal at varying temperatures.

❖ Indium-111, a radioactive isotope of indium, is used in medical imaging, particularly in single-photon emission computed tomography (SPECT) scans, to help visualise biological processes.

❖ Indium is used in the production of certain types of missiles and other military equipment due to its unique properties.

❖ The largest producers of indium are China, Canada, and Japan. It is primarily extracted as a by-product of zinc refining.

PHYSICAL PROPERTIES:

Physical Property	Description
Atomic Symbol	In
Atomic Number	49
Atomic Mass	114.82 amu
State at Room Temp	Solid (soft, silvery-white metal)
Color	Silvery-white
Odor	Odorless
Density	7.31 g/cm³
Melting Point	156.6°C
Boiling Point	2,072°C
Solubility in Water	Insoluble
Conductivity	Good conductor of electricity and heat
Abundance in Earth	Relatively rare, primarily obtained from zinc ores

CHEMICAL PROPERTIES:

Chemical Property	Description
Reactivity	Generally unreactive, but can oxidize in air
Oxidation States	Common oxidation states are +1 and +3
Electronegativity	1.78 (Pauling scale)
Reaction with Oxygen	Forms indium oxide (In_2O_3) when heated in the presence of oxygen
Reaction with Acids	Dissolves in acids, producing indium salts
Reaction with Water	Stable in water under normal conditions
Reaction with Halogens	Reacts with halogens to form indium halides
Flammability	Non-flammable but can produce toxic fumes when heated
Toxicity	Generally low toxicity; however, indium compounds can be harmful in large amounts
Biological Importance	No known biological role; considered non-essential for living organisms
Isotopes	Indium has two stable isotopes: Indium-113 and Indium-115

TIN

Hi builders! I'm Tin, a classic in soldering and alloys like bronze. From ancient times to modern tech, I'm the metal that keeps things sticking together.

INTERESTING FACTS ABOUT TIN

❖Tin has been known to humanity for thousands of years, with evidence of its use dating back to around 3000 BC. It was often used to make bronze when alloyed with copper.

❖The symbol for tin, Sn, comes from the Latin word "stannum," which means tin. This reflects its historical significance in metallurgy.

❖Tin has a relatively low melting point of 231.9°C, making it easy to work with in various applications, such as soldering and casting.

❖ Tin is considered non-toxic and safe for use in food packaging and storage, making it a popular choice for cans and food containers.

❖ Tin is commonly used to create alloys, including bronze (with copper) and pewter (with copper, antimony, and lead). These alloys have unique properties that make them valuable for various applications.

❖ Tin is widely used in soldering, particularly in electronics. Tin-lead solders have been a standard for many years, although lead-free solders are becoming more common due to health concerns.

❖ Tin has excellent resistance to corrosion, which is why it is often used to coat other metals to protect them from rust and degradation.

❖ A phenomenon known as "tin whiskers" can occur with tin-coated surfaces. These are thin, hair-like projections that can grow from tin surfaces and can lead to short circuits in electronics.

❖ Tin was used as a base metal for coins in various cultures. Its malleability and resistance to corrosion made it a suitable choice for currency.

❖ Tin compounds are used in the production of glass, especially in the form of tin oxide, which helps to improve the quality.

PHYSICAL PROPERTIES:

Physical Property	Description
Atomic Symbol	Sn
Atomic Number	50
Atomic Mass	118.71 amu
State at Room Temp	Solid (silvery-white metal)
Color	Silvery-white
Odor	Odorless
Density	7.31 g/cm³
Melting Point	231.9°C
Boiling Point	2,602°C
Solubility in Water	Insoluble
Conductivity	Fair conductor of electricity
Abundance in Earth	Relatively abundant, found in cassiterite, other ores

CHEMICAL PROPERTIES:

Chemical Property	Description
Reactivity	Relatively unreactive; does not readily react with air or water
Oxidation States	Common oxidation states are +2 and +4
Electronegativity	1.96 (Pauling scale)
Reaction with Oxygen	Forms tin oxide (SnO_2) when heated in air
Reaction with Acids	Dissolves in acids, forming tin salts
Reaction with Water	Does not react with water under normal conditions
Reaction with Halogens	Reacts with halogens to form tin halides
Flammability	Non-flammable, but finely divided tin powder can be a fire hazard
Toxicity	Generally low toxicity; however, tin compounds can be harmful in large amounts
Biological Importance	Tin is not considered essential for biological processes
Isotopes	Tin has ten stable isotopes, with Tin-120 being the most abundant

ANTIMONY

Hello flame stoppers! I'm Antimony, used in flame retardants and electronics. I might not be flashy, but I'm here to keep things safe and functional.

INTERESTING FACTS ABOUT ANTIMONY

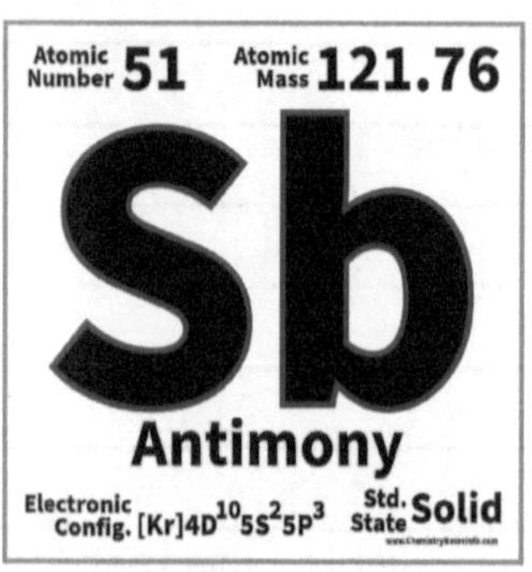

❖Antimony has a long history, dating back to ancient times. Its use was recorded as early as 1600 BC in an Egyptian papyrus. However, the element itself was first specifically studied by the French chemist Nicolas Lémery in the early 18th century. He published a detailed treatise on antimony in 1707.

❖Antimony has been known since ancient times. It was used in cosmetics in ancient Egypt and in the production of various alloys.

❖The symbol for antimony, Sb, comes from the Latin word "stibium," which means "antimony" and is derived from the Greek word "stibi," referring to the ore.

❖ Antimony exists in two allotropic forms: metallic and non-metallic. The metallic form is shiny and silver-gray, while the non-metallic form is a brittle, yellowish powder.

❖ Antimony is relatively un-reactive at room temperature but will oxidise when heated in air, forming antimony trioxide (Sb_2O_3).

❖ Antimony is commonly used in alloys to improve hardness and mechanical strength. It is often added to lead to create stronger lead-acid batteries.

❖ Antimony trioxide is used as a flame retardant in various materials, including plastics, textiles, rubber, because it inhibit combustion.

❖ Historically, antimony compounds were used in medicine to treat various ailments, including fevers and infections, although many of these uses have been discontinued due to toxicity concerns.

❖ Antimony is toxic in large amounts and can cause respiratory issues, skin irritation, and other health problems.

❖ Antimony is used in the electronics industry for producing semiconductors and as a doping agent in certain materials.

PHYSICAL PROPERTIES:

Physical Property	Description
Atomic Symbol	Sb
Atomic Number	51
Atomic Mass	121.76 amu
State at Room Temp	Solid (silver-gray, brittle metal)
Color	Silvery-gray
Odor	Odorless
Density	6.697 g/cm^3
Melting Point	630.6°C
Boiling Point	1,587°C
Solubility in Water	Insoluble
Conductivity	Poor conductor of electricity
Abundance in Earth	Relatively rare, found primarily in stibnite and other sulfide ores

CHEMICAL PROPERTIES:

Chemical Property	Description
Reactivity	Relatively unreactive at room temperature; reacts with halogens and oxygen at elevated temperatures
Oxidation States	Common oxidation states are -3, +3, and +5
Electronegativity	2.05 (Pauling scale)
Reaction with Oxygen	Forms antimony trioxide (Sb_2O_3) when heated in air
Reaction with Acids	Dissolves in concentrated sulfuric acid, forming antimony sulfate
Reaction with Water	Does not react with water under normal conditions
Reaction with Halogens	Reacts with halogens to form antimony halides
Flammability	Non-flammable but can produce toxic fumes when heated
Toxicity	Toxic; can cause respiratory issues and other health problems
Biological Importance	No known essential biological role; generally considered harmful
Isotopes	Antimony has one stable isotope: Antimony-121

TELLURIUM

Hi tech trailblazers! I'm Tellurium, essential in solar panels and alloys. I'm quietly helping shape the future of renewable energy and advanced materials.

INTERESTING FACTS ABOUT TELLURIUM

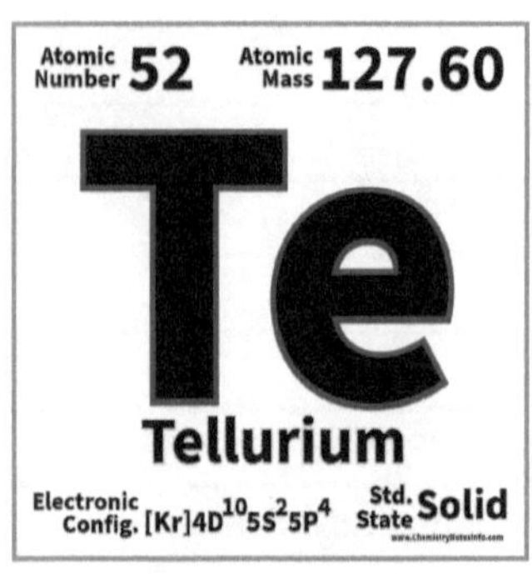

❖Tellurium is a relatively rare element in the Earth's crust, found in about 0.001% abundance. It is typically extracted from copper and lead ores.

❖Tellurium was discovered in 1782 by the Austrian chemist Franz-Joseph Müller von Reichenstein while he was studying gold ores. The name "tellurium" is derived from the Latin word "tellus," meaning "earth."

❖Tellurium is classified as a metalloid, which means it has properties of both metals and nonmetals. It is brittle and has a metallic luster.

❖ Tellurium is a semiconductor, making it useful in electronics and photovoltaic cells. It has applications in solar panels and thermoelectric devices.

❖ Tellurium is added to certain alloys to improve machinability and resistance to corrosion. It is often alloyed with lead and copper.

❖ Tellurium exhibits behavior similar to that of sulfur and selenium, forming a variety of compounds, including tellurides and oxides.

❖ Tellurium is classified as one of the rare earth metals due to its scarcity and unique properties. It is sometimes used as a dopant in rare-earth materials.

❖ Tellurium is not considered essential for biological processes, but trace amounts can be found in some organisms. It can be toxic in larger quantities.

❖ One of the most common compounds of tellurium is tellurium dioxide (TeO_2), which is used in glass production and as a catalyst in certain chemical reactions.

❖ Tellurium is used in the production of certain types of semiconductors and in thermoelectric materials, which convert temperature differences into electric voltage.

PHYSICAL PROPERTIES:

Physical Property	Description
Atomic Symbol	Te
Atomic Number	52
Atomic Mass	127.60 amu
State at Room Temp	Solid (silvery-white, brittle metal)
Color	Silvery-white
Odor	Odorless
Density	6.24 g/cm^3
Melting Point	449.5°C
Boiling Point	988°C
Solubility in Water	Insoluble
Conductivity	Semiconductor; conducts electricity better than insulators but not as well as metals
Abundance in Earth	Relatively rare, primarily obtained from copper and lead ores

CHEMICAL PROPERTIES:

Chemical Property	Description
Reactivity	Reacts with oxygen and halogens; relatively stable under normal conditions
Oxidation States	Common oxidation states are -2, +2, +4, and +6
Electronegativity	2.1 (Pauling scale)
Reaction with Oxygen	Forms tellurium dioxide (TeO_2) when heated in air
Reaction with Acids	Reacts with strong acids to form tellurium salts
Reaction with Water	Does not react with water under normal conditions
Reaction with Halogens	Reacts with halogens to form tellurium halides
Flammability	Non-flammable but can produce toxic fumes when heated
Toxicity	Can be toxic in high doses; can cause respiratory issues and other health problems
Biological Importance	No known essential biological role; considered harmful in large amounts

IODINE

Hello healers! I'm Iodine, crucial for thyroid health and medical antiseptics. From salt to surgery, I'm here to keep you healthy and safe.

INTERESTING FACTS ABOUT IODINE

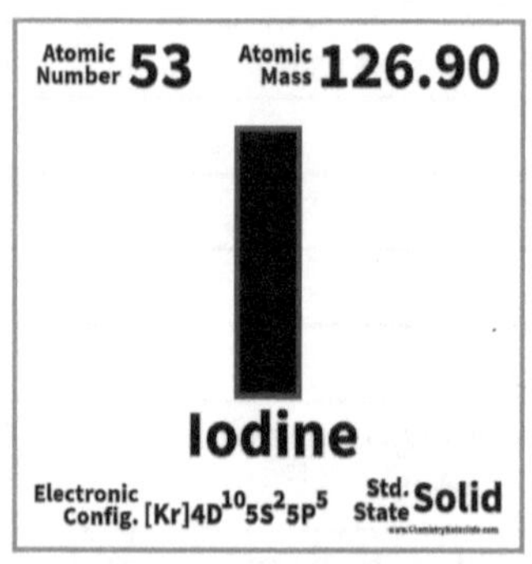

❖Iodine was discovered in 1811 by the French chemist Bernard Courtois while he was extracting sodium from seaweed. He noticed a violet vapour during the process, which turned out to be iodine.

❖Iodine exists as a diatomic molecule (I_2) and appears as dark purple-black crystals or a purple vapour when heated.

❖Iodine is an essential trace element for human health, crucial for the production of thyroid hormones. A deficiency can lead to thyroid-related diseases, such as goiter and hypothyroidism.

❖ Iodine has strong antiseptic properties and is widely used in medical applications as a disinfectant for wounds and surgical scrubs (in the form of iodine tincture).

❖ Iodine is primarily found in seawater, seaweeds, and certain fish. Foods such as dairy products, eggs, and iodised salt are also important dietary sources.

❖ When iodine is added to starch, it forms a deep blue complex, making it a useful indicator in various chemical reactions.

❖ Iodine is volatile; it readily sublimates from solid to gas at room temperature, producing a characteristic purple vapour.

❖ Iodine-131 is a radioactive isotope of iodine used in medical imaging and treatment, particularly in thyroid cancer therapy.

❖ Iodine compounds have historically been used in photography, particularly in early days of the field as part of developing agents.

❖ Iodine is used as an alloying agent in certain metals to improve their mechanical properties.

❖ Iodine deficiency is a major public health issue in many parts of the world, leading to efforts to iodise salt and other food products to improve dietary intake.

PHYSICAL PROPERTIES:

Physical Property	Description
Atomic Symbol	I
Atomic Number	53
Atomic Mass	126.90 amu
State at Room Temp	Solid (purple-black crystalline solid)
Color	Dark purple-black
Odor	Characteristic sharp odor
Density	4.93 g/cm^3
Melting Point	113.7°C
Boiling Point	184.3°C
Solubility in Water	Slightly soluble (more soluble in organic solvents)
Conductivity	Poor conductor of electricity
Abundance in Earth	Relatively rare, found in seawater and certain minerals

CHEMICAL PROPERTIES:

Chemical Property	Description
Reactivity	Reactive, especially with metals and nonmetals
Oxidation States	Common oxidation states are -1, +1, +5, and +7
Electronegativity	2.66 (Pauling scale)
Reaction with Oxygen	Can form iodine oxides, such as I_2O and IO_2
Reaction with Acids	Reacts with strong acids to form iodine salts
Reaction with Water	Slightly soluble in water, forming hydroiodic acid
Reaction with Halogens	Reacts with halogens to form interhalogen compounds
Flammability	Non-flammable but can produce toxic vapors when heated
Toxicity	Can be toxic in high doses; inhalation of iodine vapor can cause respiratory issues
Biological Importance	Essential for thyroid hormone production; deficiency leads to various health issues
Isotopes	Iodine has one stable isotope: Iodine-127; Iodine-131 is a common radioactive isotope used in medicine

XENON

Hi bright lights! I'm Xenon, shining in car headlights and high-powered lamps. You'll find me lighting up the night with my intense glow.

INTERESTING FACTS ABOUT XENON

❖Xenon is a member of the noble gases group (Group 18) in the periodic table, known for its lack of chemical reactivity due to having a full valence shell of electrons.

❖Xenon was discovered in 1898 by the Scottish chemist William Ramsay and his assistant Morris Travers while they were studying the products of liquid air. They isolated it from other gases.

❖Xenon is a colourless, odourless, and tasteless gas at room temperature, making it difficult to detect without specialised equipment.

❖ It is one of the heaviest noble gases, with a density nearly five times that of air, making it useful in certain applications like lighting.

❖ Xenon emits a beautiful blue glow when electrified, which is why it is often used in high-intensity discharge lamps, including those used in film projectors and vehicle headlights.

❖ Xenon is used in medical imaging and anaesthetic applications due to its anaesthetic properties and ability to provide high-quality imaging in CT scans.

❖ With an atomic mass of 131.29 amu, xenon is one of the heaviest stable gases known, making it unique among the noble gases.

❖ Although noble gases are typically inert, xenon can form compounds with fluorine and oxygen, including xenon difluoride (XeF_2) and xenon tetrafluoride (XeF_4).

❖ Xenon is used in certain types of ion propulsion systems for spacecraft due to its high atomic mass and low reactivity, which enhance efficiency.

❖ Xenon flash lamps are used in high-speed photography and scientific instruments due to their ability to produce very bright and short bursts of light.

PHYSICAL PROPERTIES:

Physical Property	Description
Atomic Symbol	Xe
Atomic Number	54
Atomic Mass	131.29 amu
State at Room Temp	Gas (colorless, odorless)
Color	Colorless
Odor	Odorless
Density	5.897 g/L (at 0°C and 1 atm)
Melting Point	-111.8°C
Boiling Point	-108.1°C
Solubility in Water	Slightly soluble
Conductivity	Non-conductor of electricity
Abundance in Earth	Present in trace amounts in the atmosphere

CHEMICAL PROPERTIES:

Chemical Property	Description
Reactivity	Very low reactivity; forms compounds only under specific conditions
Oxidation States	Common oxidation states are +2, +4, +6
Electronegativity	Not defined, but inert behavior is observed
Reaction with Oxygen	Can form xenon oxides, such as XeO_3 and XeO_2
Reaction with Halogens	Can react with halogens to form xenon halides
Flammability	Non-flammable
Toxicity	Generally considered safe; inert and non-toxic under normal conditions
Biological Importance	No known essential biological role; used in some medical applications
Isotopes	Xenon has several stable isotopes, with Xenon-132 being the most abundant

CESIUM

Hello timekeepers! I'm Cesium, the element behind the most accurate atomic clocks. Precision is my game, and I'm the master of keeping perfect time.

INTERESTING FACTS ABOUT CESIUM

❖Cesium is a soft, gold-coloured metal that is so soft that it can be cut with a knife at room temperature.

❖Cesium was discovered in 1860 by the German chemist Robert Bunsen and his assistant Gustav Kirchhoff using a spectroscope while analysing mineral water.

❖Cesium has one of the lowest melting points of any metal, melting at 28.5°C (83.3°F), which means it can be liquid on a warm day.

❖Cesium is extremely reactive, particularly with water, producing hydrogen gas and cesium hydroxide. This reactivity requires cesium to be stored under oil to prevent reactions with moisture.

❖ Cesium is most famous for its role in atomic clocks, which use the vibrations of cesium atoms to keep time with incredible precision. The definition of the second is based on cesium's vibrations.

❖ Cesium is found in nature mainly in the mineral pollucite, which is mined primarily in Canada and Namibia.

❖ While cesium is relatively rare in the Earth's crust, it is more abundant in the universe. It is often obtained as a byproduct of lithium and potassium extraction.

❖ Cesium forms a variety of compounds, most notably cesium chloride (CsCl), which is used in various applications, including as a contrast agent in medical imaging.

❖ Cesium has no known biological role in humans and can be toxic in large quantities, but it is used in certain medical treatments.

❖ Cesium has the atomic mass of approximately 132.91 amu, making it one of the heaviest stable alkali metals.

❖ Cesium formate is used in drilling fluids due to its high density and ability to control wellbore stability.

PHYSICAL PROPERTIES:

Physical Property	Description
Atomic Symbol	Cs
Atomic Number	55
Atomic Mass	132.91 amu
State at Room Temp	Solid (soft, metallic)
Color	Silvery-gold
Odor	Odorless
Density	1.93 g/cm^3
Melting Point	28.5°C
Boiling Point	671°C
Solubility in Water	Highly soluble
Conductivity	Good conductor of electricity
Abundance in Earth	Rare, found primarily in minerals like pollucite and lepidolite

CHEMICAL PROPERTIES:

Chemical Property	Description
Reactivity	Highly reactive, especially with water and air
Oxidation States	Common oxidation states are +1
Electronegativity	0.79 (Pauling scale)
Reaction with Water	Reacts vigorously with water to form cesium hydroxide and hydrogen gas
Reaction with Acids	Reacts with acids to form cesium salts
Reaction with Oxygen	Burns in air to form cesium oxide (Cs_2O)
Flammability	Highly flammable; must be handled with caution
Toxicity	Can be toxic in high doses; not essential for human health
Biological Importance	No known essential biological role; can be harmful in large amounts
Isotopes	Cesium has one stable isotope: Cesium-133, widely used in atomic clocks

BARIUM

Hi medical imagers! I'm Barium, used in X-ray imaging to make the inside of your body visible. I also bring green fireworks to life with my vivid colour.

INTERESTING FACTS ABOUT BARIUM

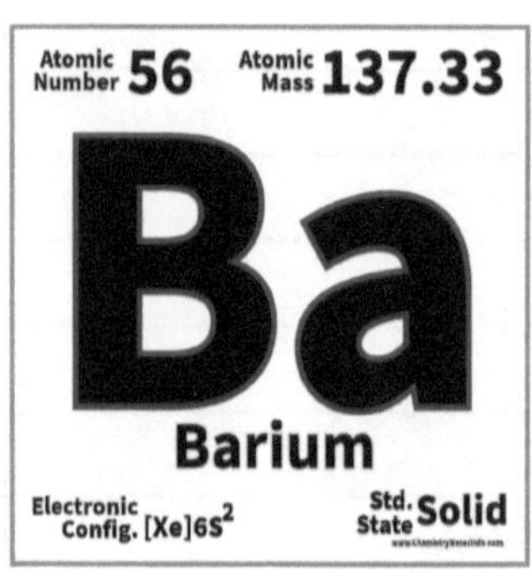

❖Barium is a soft, silvery-white alkaline earth metal that is so soft it can be cut with a knife.

❖Barium was discovered in 1774 by the Swedish chemist Carl Wilhelm Scheele, who isolated it from barite (barium sulfate) while trying to identify the metal contained within.

❖Barium belongs to the alkaline earth metals group (Group 2) in the periodic table, which are known for their reactivity, especially with water.

❖Barium is one of the heaviest alkaline earth metals, with a density of about 3.62 g/cm³, making it much denser than many common metals.

❖ Barium reacts with water to produce barium hydroxide and hydrogen gas. This reaction is vigorous, and barium must be stored under oil to prevent it from reacting with moisture in the air.

❖ Barium sulfate is commonly used in medical imaging, particularly in X-ray examinations of the gastrointestinal tract, as it is opaque to X-rays and allows for clear imaging.

❖ Barium compounds are used in fireworks to produce green colors, making it a popular choice for pyrotechnics.

❖ Barium is used to improve the quality and properties of certain types of glass, such as in the production of glass for cathode ray tubes and other electronic applications.

❖ Barium alloys are used in vacuum tubes to remove remaining gases (a process called "gettering"), which improves their efficiency and extends their lifespan.

❖ Barium is naturally found in various minerals, with barite ($BaSO_4$) and witherite ($BaCO_3$) being the most common sources of barium.

❖ Barium can form a variety of compounds, including barium oxide (BaO), barium chloride ($BaCl_2$), and barium carbonate ($BaCO_3$), which are used in various industrial applications.

PHYSICAL PROPERTIES:

Physical Property	Description
Atomic Symbol	Ba
Atomic Number	56
Atomic Mass	137.33 amu
State at Room Temp	Solid (soft, metallic)
Color	Silvery-white
Odor	Odorless
Density	3.62 g/cm^3
Melting Point	727°C
Boiling Point	1640°C
Solubility in Water	Moderately soluble (barium hydroxide is soluble)
Conductivity	Good conductor of electricity
Abundance in Earth	Commonly found in the Earth's crust, primarily in minerals like barite and witherite

CHEMICAL PROPERTIES:

Chemical Property	Description
Reactivity	Highly reactive, especially with water and acids
Oxidation States	Common oxidation states are +2
Electronegativity	0.89 (Pauling scale)
Reaction with Water	Reacts with water to produce barium hydroxide and hydrogen gas
Reaction with Acids	Reacts with acids to form barium salts
Reaction with Oxygen	Burns in air to form barium oxide (BaO)
Flammability	Flammable when powdered or in thin strips
Toxicity	Toxic in large amounts; can cause barium poisoning
Biological Importance	No known essential biological role; can be harmful in high doses
Isotopes	Barium has several stable isotopes, with Barium-138 being the most abundant

LANTHANUM

Hello optics experts! I'm Lanthanum, improving the quality of camera lenses and catalysts. From science to photography, I help the world see more clearly.

INTERESTING FACTS ABOUT LANTHANUM

❖Lanthanum is the first element in the lanthanide series, which is a group of 15 rare earth elements from atomic numbers 57 to 71.

❖Lanthanum is a soft, malleable, silvery-white metal that tarnishes when exposed to air.

❖Lanthanum was discovered in 1839 by Swedish chemist Carl Gustaf Mosander while he was analysing cerium oxide. The name "lanthanum" comes from the Greek word "lanthanein," meaning "to lie hidden."

❖ Although considered a rare earth element, lanthanum is not as rare as the name implies. It is relatively abundant in the Earth's crust, even more so than lead.

❖ Lanthanum is a critical component in the nickel-metal hydride (NiMH) batteries used in hybrid vehicles such as the Toyota Prius.

❖ Lanthanum is used in the manufacture of catalytic converters in automobiles to reduce toxic emissions.

❖ Lanthanum oxide is used in the production of high-quality optical glass, such as camera lenses and telescope lenses, due to its high refractive index.

❖ It is used in carbon arc lamps, which were once commonly used in the film industry for studio lighting.

❖ When exposed to air, lanthanum quickly forms an oxide layer, which protects the metal underneath from further corrosion.

❖ Lanthanum is used in the production of high-performance spark plug electrodes for engines due to its conductive and heat-resistant properties.

❖ Lanthanum has a few radioactive isotopes, including Lanthanum-138, which is used in research and geological dating.

PHYSICAL PROPERTIES:

Physical Property	Description
Atomic Symbol	La
Atomic Number	57
Atomic Mass	138.91 amu
State at Room Temp	Solid (soft, metallic)
Color	Silvery-white
Odor	Odorless
Density	6.146 g/cm^3
Melting Point	920°C
Boiling Point	3464°C
Solubility in Water	Insoluble
Conductivity	Good conductor of heat and electricity
Abundance in Earth	Found in minerals such as monazite and bastnäsite, relatively abundant

CHEMICAL PROPERTIES:

Chemical Property	Description
Reactivity	Reacts slowly with cold water, rapidly with hot water, and forms lanthanum hydroxide
Oxidation States	+3 is the most common oxidation state
Electronegativity	1.1 (Pauling scale)
Reaction with Oxygen	Reacts with oxygen to form lanthanum oxide (La_2O_3)
Reaction with Acids	Reacts with most acids to produce lanthanum salts
Flammability	Not flammable under normal conditions but can ignite when finely divided
Toxicity	Generally considered non-toxic, but lanthanum compounds can be harmful if inhaled or ingested
Biological Importance	No known biological role
Isotopes	Lanthanum has one stable isotope, Lanthanum-139, and a few radioactive isotopes

CERIUM

Hi innovators! I'm Cerium, a key part of catalytic converters and glass polishing. Whether reducing pollution or perfecting optics, I'm here to innovate.

INTERESTING FACTS ABOUT CERIUM

❖Cerium is the most abundant of all the rare earth elements and is more common in the Earth's crust than copper.

❖Cerium was named after the dwarf planet Ceres, which was discovered just two years before cerium itself, in 1801.

❖Cerium was discovered by Swedish chemists Jöns Jakob Berzelius and Wilhelm Hisinger, and independently by Martin H. Klaproth, from a mineral called cerite.

❖Cerium is highly reactive with air, forming a dark oxide layer on its surface when exposed, making it appear tarnished.

❖ Cerium oxide (CeO_2) is used in catalytic converters to help reduce harmful emissions from vehicle exhaust systems.

❖ Cerium is a key component in lighter flints. A mixture of cerium and iron forms ferrocerium, which sparks when struck.

❖ Cerium oxide is used as a polishing compound for glass and semiconductors. It's commonly used to polish telescope lenses, mirrors, and even automobile windows.

❖ Cerium oxide is used in coatings that allow ovens to self-clean, helping to break down food residues when heated to high temperatures.

❖ Due to its high reactivity, cerium can spontaneously ignite when scratched or cut, especially in finely divided forms.

❖ Cerium is used in various alloys to improve strength and resistance to heat and corrosion. For example, it is added to aluminum alloys to make them more heat-resistant.

❖ Cerium plays a role in renewable energy research, as its compounds are being tested in solar panels and water-splitting reactions to produce clean hydrogen fuel.

❖ Cerium compounds are being researched for use in high-temperature fuel cells, where they can enhance conductivity and efficiency.

PHYSICAL PROPERTIES:

Physical Property	Description
Atomic Symbol	Ce
Atomic Number	58
Atomic Mass	140.12 amu
State at Room Temp	Solid (soft, metallic)
Color	Silvery-white, but tarnishes to a grayish color when exposed to air
Odor	Odorless
Density	6.770 g/cm^3
Melting Point	798°C
Boiling Point	3360°C
Solubility in Water	Insoluble
Conductivity	Good conductor of heat and electricity
Abundance in Earth	Found in minerals like monazite and bastnäsite, relatively abundant for a rare earth element

CHEMICAL PROPERTIES:

Chemical Property	Description
Reactivity	Reacts readily with air and water, forming cerium(III) oxide (Ce_2O_3) or cerium(IV) oxide (CeO_2)
Oxidation States	+3 and +4
Electronegativity	1.12 (Pauling scale)
Reaction with Water	Slowly reacts with water, forming cerium hydroxide and hydrogen gas
Reaction with Acids	Dissolves in acids to form cerium salts
Reaction with Oxygen	Cerium reacts easily with oxygen to form cerium oxide (CeO_2), which is its most common oxide
Flammability	Highly flammable when finely divided
Toxicity	Generally considered non-toxic, though cerium compounds can be hazardous in large amounts
Biological Importance	No known biological role; used in some experimental medical applications
Isotopes	Cerium has four naturally occurring isotopes, with Cerium-140 being the most abundant

PRASEODYMIUM

Hello magnet makers! I'm Praseodymium, helping to create strong magnets used in wind turbines and electric cars. I'm all about sustainable power!

INTERESTING FACTS ABOUT PRASEODYMIUM

❖The name "praseodymium" comes from the Greek words "prasios didymos," meaning "green twin," due to the green salts it forms.

❖Praseodymium was discovered in 1885 by Austrian chemist Carl Auer von Welsbach, who separated it from neodymium.

❖Praseodymium forms bright green-colored oxides, which are responsible for its name and give it unique properties in glass-making.

❖Praseodymium is used to color glass and ceramics, producing deep yellow shades. It is commonly used in glassblowers' goggles to protect eyes from intense light.

❖ Praseodymium is used in the creation of high-strength permanent magnets, particularly in alloys with neodymium for use in motors, headphones, and other devices.

❖ It is sometimes used in catalytic converters in automobiles to help reduce harmful emissions.

❖ Praseodymium is added to alloys used in aircraft engines to increase the strength and heat resistance of the materials.

❖ Praseodymium ions are used in lasers, particularly in fiber-optic communication, because of their ability to emit sharp wavelengths.

❖ Praseodymium is classified as a rare earth element but is relatively abundant in the Earth's crust compared to other rare earth metals.

❖ Praseodymium is a component of mischmetal, an alloy used in flints for lighters and other pyrophoric devices.

❖ Praseodymium oxide is used to create a bright yellow color in glass and enamel, which is highly valued in ceramics and art glass.

❖ When alloyed with magnesium, praseodymium improves corrosion resistance of metal, making it useful for high-strength applications.

PHYSICAL PROPERTIES:

Physical Property	Description
Atomic Symbol	Pr
Atomic Number	59
Atomic Mass	140.91 amu
State at Room Temp	Solid (soft, metallic)
Color	Silvery-white
Odor	Odorless
Density	6.77 g/cm^3
Melting Point	931°C
Boiling Point	3520°C
Solubility in Water	Insoluble
Conductivity	Good conductor of heat and electricity
Abundance in Earth	Found in minerals like monazite and bastnäsite, relatively abundant for a rare earth element

CHEMICAL PROPERTIES:

Chemical Property	Description
Reactivity	Praseodymium reacts slowly with oxygen, forming a green oxide layer (Pr_2O_3) on its surface
Oxidation States	+3 (most stable), +4 in some compounds
Electronegativity	1.13 (Pauling scale)
Reaction with Water	Slowly reacts with water, forming praseodymium hydroxide and hydrogen
Reaction with Acids	Dissolves in acids to form praseodymium salts
Reaction with Oxygen	Reacts with oxygen to form praseodymium oxide (Pr_2O_3)
Flammability	Can be flammable when finely divided
Toxicity	Considered mildly toxic, but praseodymium compounds should be handled with care
Biological Importance	No known biological role
Isotopes	Praseodymium has only one stable isotope: Praseodymium-141

NEODYMIUM

Hey energy savers! I'm Neodymium, used in powerful magnets for motors and speakers. I keep your gadgets running efficiently and smoothly.

INTERESTING FACTS ABOUT NEODYMIUM

❖The name "neodymium" comes from the Greek words "neos didymos," meaning "new twin," as it was originally thought to be part of the element didymium before being separated.

❖Neodymium was discovered in 1885 by Austrian chemist Carl Auer von Welsbach, who separated it from praseodymium.

❖Neodymium is most famously used in the production of the neodymium magnets ($Nd_2Fe_{14}B$), which are the strongest permanent magnets available.

❖ Neodymium magnets are widely used in high-performance motors, headphones, microphones, hard disk drives, and wind turbines.

❖ Neodymium is used to color glass and ceramics, producing vibrant violet and reddish-purple hues. Neodymium glass is used in sunglasses, particularly for welders, and in laser glasses.

❖ Neodymium-doped crystals are used in Nd lasers (Neodymium-doped Yttrium Aluminum Garnet lasers), which are used in medical procedures, industrial cutting, and laser pointers.

❖ Neodymium is critical in green energy technologies like electric vehicles and wind turbines because of its role in high-strength, lightweight magnets.

❖ Although classified as a rare earth element, neodymium is relatively abundant in the Earth's crust and is found in minerals such as monazite and bastnäsite.

❖ Neodymium is used in "dichroic" glass, which changes color depending on the lighting, appearing purple in sunlight and blue in artificial light.

❖ Neodymium magnets are strong but lose magnetism at high temperatures, which limits their use in some high-heat environments. Research is ongoing to improve this property.

PHYSICAL PROPERTIES:

Physical Property	Description
Atomic Symbol	Nd
Atomic Number	60
Atomic Mass	144.24 amu
State at Room Temp	Solid (soft, metallic)
Color	Silvery-white but tarnishes to yellowish or purple when exposed to air
Odor	Odorless
Density	7.01 g/cm^3
Melting Point	1024°C
Boiling Point	3074°C
Solubility in Water	Insoluble
Conductivity	Good conductor of heat and electricity
Abundance in Earth	Found in minerals like monazite and bastnäsite, relatively abundant for a rare earth element

CHEMICAL PROPERTIES:

Chemical Property	Description
Reactivity	Reacts with oxygen and air, forming neodymium oxide (Nd_2O_3)
Oxidation States	+3 (most common), +2 in some compounds
Electronegativity	1.14 (Pauling scale)
Reaction with Water	Reacts slowly with cold water, more rapidly with hot water, forming neodymium hydroxide
Reaction with Acids	Dissolves in acids to form neodymium salts
Reaction with Oxygen	Reacts quickly with oxygen to form an oxide layer
Flammability	Flammable in finely divided form
Toxicity	Considered mildly toxic, should be handled with care
Biological Importance	No known biological role
Isotopes	Neodymium has seven naturally occurring isotopes, with Neodymium-142 being the most abundant

PROMETHIUM

Hello glow seekers! I'm Promethium, a rare radioactive element found in some glow-in-the-dark materials and nuclear batteries. I light up the night!

INTERESTING FACTS ABOUT PROMETHIUM

❖Promethium is named after Prometheus, a Titan from Greek mythology who stole fire from the gods and gave it to humanity. This reflects the element's radioactive and energetic nature.

❖Promethium was discovered in 1945 by American chemists Jacob Marinsky and Lawrence Glendenin while studying the byproducts of uranium fission at the Oak Ridge National Laboratory.

❖Promethium is unique among the lanthanides because it is the only one that is entirely radioactive, with no stable isotopes.

❖ Promethium is used in nuclear batteries, particularly in devices like pacemakers and space probes, where its radioactive decay provides a steady source of power.

❖ Promethium compounds can glow in the dark, making them useful for luminous paint, dials, and instrument panels, though their use is limited due to radioactivity concerns.

❖ Promethium is extremely rare in nature. It is mostly found in trace amounts in uranium ores or produced artificially in nuclear reactors.

❖ The most stable isotope of promethium is Pm-145, which has a half-life of 17.7 years. Other isotopes have much shorter half-lives, making the element challenging to work with.

❖ Promethium has no known biological role and is considered highly toxic due to its radioactivity.

❖ Trace amounts of promethium have been detected in the spectra of some stars, suggesting that it is produced during the stellar nucleosynthesis process.

❖ Early predictions of promethium's existence were made by Dmitri Mendeleev, but its actual discovery was delayed due to its scarcity and radioactivity, which made it difficult to isolate.

PHYSICAL PROPERTIES:

Physical Property	Description
Atomic Symbol	Pm
Atomic Number	61
Atomic Mass	[145] (no stable isotopes)
State at Room Temp	Solid (metallic, highly radioactive)
Color	Silvery-white or metallic, but tarnishes easily
Odor	Odorless
Density	7.26 g/cm^3
Melting Point	1100°C
Boiling Point	3000°C (estimated)
Solubility in Water	Insoluble
Conductivity	Good conductor of heat and electricity
Abundance in Earth	Extremely rare, mostly found as a byproduct of uranium fission

CHEMICAL PROPERTIES:

Chemical Property	Description
Reactivity	Promethium reacts with oxygen to form promethium oxide (Pm_2O_3), and with acids to form salts
Oxidation States	+3 is the most stable oxidation state
Electronegativity	1.13 (Pauling scale)
Radioactivity	All isotopes of promethium are radioactive, with Pm-145 being the most stable
Reaction with Water	Reacts slowly with water, forming promethium hydroxide
Reaction with Acids	Dissolves in acids to form promethium salts
Flammability	Not flammable, but radioactive material can pose safety hazards
Toxicity	Highly toxic due to radioactivity; must be handled with care
Biological Importance	No known biological role
Isotopes	No stable isotopes; Pm-145 (half-life 17.7 years) is the most common isotope

SAMARIUM

Hi strong magnets! I'm Samarium, powering some of the strongest magnets in the world. You'll find me in headphones, electric motors, and space tech.

INTERESTING FACTS ABOUT SAMARIUM

❖Samarium is named after the mineral "samarskite," which was in turn named after a Russian mining engineer, Colonel Vasili Samarsky-Bykhovets.

❖Samarium was discovered in 1879 by French chemist Paul-Émile Lecoq de Boisbaudran, who isolated it from the mineral samarskite.

❖Samarium was one of the first rare earth elements to be discovered and isolated, leading to the identification of other lanthanides.

❖ Samarium is an important component in samarium-cobalt (SmCo) magnets, which are known for their ability to maintain magnetism at high temperatures and resist demagnetisation.

❖ The radioactive isotope Samarium-153 is used in cancer treatments to relieve pain in patients with bone cancer. It is administered in small doses and helps target and destroy cancerous cells.

❖ Samarium oxide is used in optical glass and infrared-absorbing glass, which protects against heat and ultraviolet light.

❖ Samarium is used as a neutron absorber in nuclear reactors to help control the fission process. It is a byproduct of nuclear reactions and contributes to the stability of the reactor.

❖ Samarium is used in lasers and masers (microwave amplification by stimulated emission of radiation), where it helps amplify signals at microwave frequencies. Samarium is also detected in stars and Sun, and its spectral lines are used in astrophysics to study stellar composition and the processes of nucleosynthesis in the universe.

❖ Samarium is used in the phosphor coatings of fluorescent lamps, enhancing their brightness and colour.

❖ While samarium has no known biological role, some of its compounds are being researched for potential applications in medicine due to their interaction with radiation.

PHYSICAL PROPERTIES:

Physical Property	Description
Atomic Symbol	Sm
Atomic Number	62
Atomic Mass	150.36 amu
State at Room Temp	Solid (soft, metallic)
Color	Silvery-white
Odor	Odorless
Density	7.52 g/cm^3
Melting Point	1072°C
Boiling Point	1794°C
Solubility in Water	Insoluble
Conductivity	Good conductor of heat and electricity
Abundance in Earth	Found in minerals like monazite and bastnäsite, relatively abundant for a rare earth element

CHEMICAL PROPERTIES:

Chemical Property	Description
Reactivity	Samarium reacts slowly with oxygen to form a protective oxide layer (Sm_2O_3)
Oxidation States	+2 and +3 (with +3 being the most stable)
Electronegativity	1.17 (Pauling scale)
Reaction with Water	Reacts slowly with water, forming samarium hydroxide
Reaction with Acids	Dissolves in acids to form samarium salts
Reaction with Oxygen	Forms samarium oxide (Sm_2O_3) when exposed to air
Flammability	Flammable in finely divided form
Toxicity	Considered mildly toxic, though generally not harmful in small amounts
Biological Importance	No known biological role
Isotopes	Samarium has several isotopes, with Samarium-152 being the most stable

EUROPIUM

Hello bright screens! I'm Europium, responsible for the red and blue colours in your TV and computer screens. I add a splash of colour to modern life.

INTERESTING FACTS ABOUT EUROPIUM

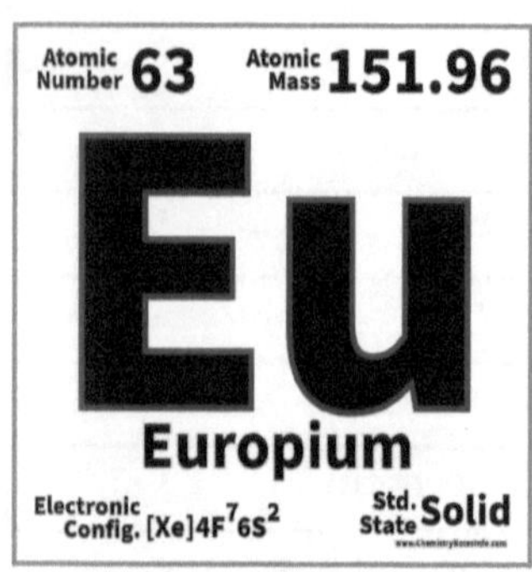

❖Europium was named after the continent of Europe, reflecting its discovery by a European scientist.

❖Europium was discovered in 1901 by French chemist Eugène-Anatole Demarçay, who identified it while studying the spectral lines of rare earth elements.

❖Europium is the most reactive of all the rare earth elements (lanthanides). It oxidises quickly in air and reacts readily with water.

❖Europium is famous for its use in red phosphor coatings in cathode ray tubes (CRTs) for colour television screens and in modern LED displays. Its compound europium(III) oxide gives off a bright red colour.

❖ Europium-doped phosphors are used in the anti-counterfeiting measures of Euro banknotes. The fluorescence from europium compounds helps verify authenticity under UV light.

❖ Europium-151 and Europium-153 are used as neutron absorbers in nuclear reactors, helping to regulate the fission process.

❖ Europium compounds are used in luminescent paints, including glow-in-the-dark applications and fluorescent lights.

❖ While europium is a rare earth element, it is more abundant than silver in the Earth's crust. It is found in minerals like monazite and bastnäsite.

❖ Europium-doped phosphors are widely used in fluorescent and compact fluorescent lamps (CFLs), enhancing brightness and colour quality.

❖ Europium has the highest neutron capture cross-section of any element, making it especially valuable for nuclear control applications.

❖ Despite being a metal, europium is surprisingly soft—you can cut it easily with a knife.

❖ Europium is used in some medical applications, such as in certain luminescent markers for biological assays, where it helps detect specific proteins or DNA.

PHYSICAL PROPERTIES:

Physical Property	Description
Atomic Symbol	Eu
Atomic Number	63
Atomic Mass	151.96 amu
State at Room Temp	Solid (soft, metallic)
Color	Silvery-white but quickly tarnishes in air
Odor	Odorless
Density	5.24 g/cm^3
Melting Point	826°C
Boiling Point	1529°C
Solubility in Water	Insoluble
Conductivity	Moderate conductor of heat and electricity
Abundance in Earth	Found in minerals like monazite and bastnäsite, relatively abundant for a rare earth element

CHEMICAL PROPERTIES:

Chemical Property	Description
Reactivity	Europium is highly reactive, especially in air, where it forms europium oxide (Eu_2O_3)
Oxidation States	+2 and +3 (with +3 being more common)
Electronegativity	1.2 (Pauling scale)
Reaction with Water	Reacts rapidly with water, forming europium hydroxide and hydrogen gas
Reaction with Acids	Dissolves readily in acids to form europium salts
Reaction with Oxygen	Reacts with oxygen to form europium oxide (Eu_2O_3)
Flammability	Highly flammable when finely divided or powdered
Toxicity	Considered mildly toxic, though not dangerous in small quantities
Biological Importance	No known biological role
Isotopes	Europium has two naturally occurring isotopes: Europium-151 and Europium-153

GADOLINIUM

Hi MRI scanners! I'm Gadolinium, used in medical imaging to help doctors see inside the body. I'm also found in alloys that work in extreme environments.

INTERESTING FACTS ABOUT GADOLINIUM

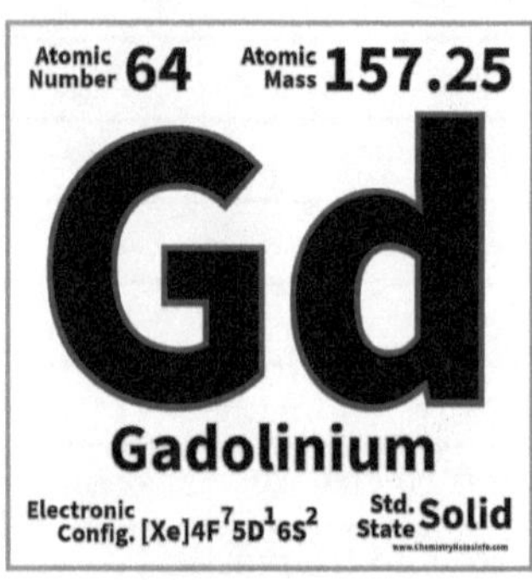

❖Gadolinium is named in honor of Finnish chemist Johan Gadolin, who was a pioneer in the study of rare earth elements.

❖Gadolinium was discovered in 1880 by Swiss chemist Jean Charles Galissard de Marignac, who found it in the mineral gadolinite.

❖Gadolinium is ferromagnetic at room temperature, meaning it is strongly attracted to magnets, but it loses this property above 20°C (68°F), becoming paramagnetic.

❖ Gadolinium is widely used as a contrast agent in magnetic resonance imaging (MRI). Its paramagnetic properties make it useful for enhancing the quality of images in medical diagnostics.

❖ Gadolinium has an excellent ability to absorb neutrons and is used as a neutron poison in nuclear reactors, helping to control the rate of fission reactions.

❖ Gadolinium-based compounds are used in neutron capture therapy, a type of cancer treatment that targets tumour cells using gadolinium's neutron absorption properties.

❖ Gadolinium exhibits the magneto-caloric effect, where its temperature changes when exposed to a magnetic field. This property is being researched for applications in magnetic refrigeration.

❖ Gadolinium is added to certain alloys to improve their resistance to oxidation and high-temperature corrosion, making them useful in aerospace applications.

❖ Gadolinium compounds, such as gadolinium oxide, are used in phosphors for fluorescent lamps and colour televisions, where they help produce high-quality light.

❖ Gadolinium is more abundant than elements like lead and is commonly extracted from minerals such as monazite & bastnäsite.

PHYSICAL PROPERTIES:

Physical Property	Description
Atomic Symbol	Gd
Atomic Number	64
Atomic Mass	157.25 amu
State at Room Temp	Solid (soft, metallic)
Color	Silvery-white
Odor	Odorless
Density	7.90 g/cm³
Melting Point	1313°C
Boiling Point	3273°C
Solubility in Water	Insoluble
Conductivity	Good conductor of heat and electricity
Abundance in Earth	Found in minerals like monazite and bastnäsite, relatively abundant for a rare earth element

CHEMICAL PROPERTIES:

Chemical Property	Description
Reactivity	Reacts with oxygen to form gadolinium oxide (Gd_2O_3) forming a protective layer
Oxidation States	+3 is the most stable oxidation state
Electronegativity	1.20 (Pauling scale)
Reaction with Water	Reacts slowly with water, forming gadolinium hydroxide
Reaction with Acids	Dissolves in most acids to form gadolinium salts
Reaction with Oxygen	Reacts readily with oxygen to form an oxide coating
Flammability	Flammable when in powdered form
Toxicity	Gadolinium compounds used in medical imaging are toxic if not properly administered
Biological Importance	No known biological role
Isotopes	Gadolinium has several isotopes, with Gadolinium-158 being the most stable

TERBIUM

Hello screen lovers! I'm Terbium, bringing vivid green colours to your displays. I'm also a key player in energy-efficient lighting.

INTERESTING FACTS ABOUT TERBIUM

❖Terbium is named after the village of Ytterby in Sweden, where several rare earth elements were discovered.

❖Terbium was discovered in 1843 by Swedish chemist Carl Gustaf Mosander while he was studying yttria, an oxide of yttrium, from the mineral gadolinite.

❖Terbium is used to produce the green phosphor in colour television tubes and fluorescent lamps. It helps create the vibrant green colour in screens and displays.

❖ Terbium compounds are also used in fluorescent lamps and compact fluorescent light bulbs (CFLs), helping them emit a bright and energy-efficient light.

❖ Terbium is used in the production of Terfenol-D, a magnetostrictive alloy that changes shape in response to a magnetic field. This alloy is used in sonar systems, actuators, and sensors.

❖ Terbium ions are known for their strong luminescence, especially in compounds like terbium oxide (Tb_2O_3), which are used in various lighting and display technologies.

❖ Like many lanthanides, Terbium can absorb neutrons and is used as a neutron absorber in nuclear reactors to help regulate the fission process.

❖ Terbium compounds are used in X-ray screens to enhance image quality and reduce exposure time, as they improve the brightness of the images.

❖ Although classified as a rare earth element, Terbium is relatively abundant in the Earth's crust and is typically extracted from minerals like monazite and bastnäsite.

❖ Terbium has no known biological role, but its compounds must be handled carefully due to potential toxicity.

PHYSICAL PROPERTIES:

Physical Property	Description
Atomic Symbol	Tb
Atomic Number	65
Atomic Mass	158.93 amu
State at Room Temp	Solid (soft, metallic)
Color	Silvery-gray
Odor	Odorless
Density	8.23 g/cm^3
Melting Point	1356°C
Boiling Point	3230°C
Solubility in Water	Insoluble
Conductivity	Good conductor of heat and electricity
Abundance in Earth	Found in minerals like monazite and bastnäsite, relatively abundant for a rare earth element

CHEMICAL PROPERTIES:

Chemical Property	Description
Reactivity	Terbium reacts slowly with air, forming a protective oxide layer (Tb_2O_3)
Oxidation States	+3 is the most common oxidation state
Electronegativity	1.2 (Pauling scale)
Reaction with Water	Reacts slowly with water, forming terbium hydroxide and hydrogen gas
Reaction with Acids	Dissolves in acids to form terbium salts
Reaction with Oxygen	Reacts with oxygen to form terbium oxide (Tb_2O_3)
Flammability	Flammable in powdered form or in finely divided states
Toxicity	Terbium compounds are considered mildly toxic if inhaled or ingested
Biological Importance	No known biological role
Isotopes	Terbium has one stable isotope, Terbium-159, and several radioactive isotopes

DYSPROSIUM

Hey tech pioneers! I'm Dysprosium, helping high-performance magnets keep their strength at high temperatures. You'll find me in electric vehicles and wind turbines.

INTERESTING FACTS ABOUT DYSPROSIUM

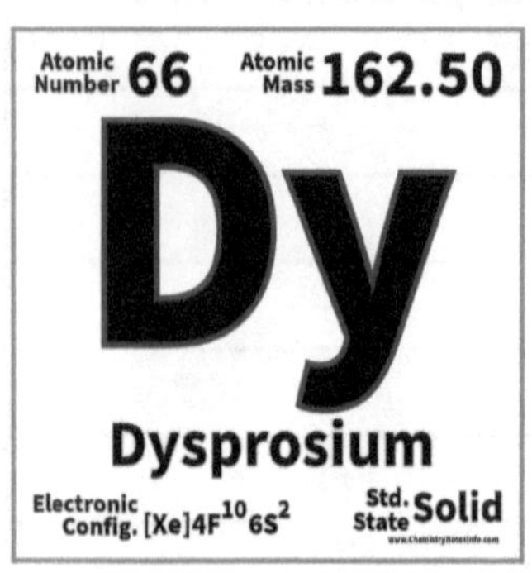

❖The name "Dysprosium" comes from the Greek word "dysprositos," meaning "hard to get," because of the difficulty in isolating the element.

❖Dysprosium was discovered in 1886 by French chemist Paul-Émile Lecoq de Boisbaudran while he was analyzing rare earth elements.

❖Dysprosium is used in the production of Terfenol-D, a magnetostrictive alloy that changes shape under magnetic fields, making it useful in sonar systems, sensors, and actuators.

❖ Dysprosium is used in nuclear reactors as a neutron absorber due to its high ability to absorb neutrons without undergoing fission.

❖ Dysprosium is used in magnets that operate at high temperatures, such as those in wind turbines, electric vehicles, and hybrid cars. It enhances the heat resistance of neodymium-iron-boron magnets.

❖ Dysprosium compounds are used in phosphorescent materials, including glow-in-the-dark paints and materials, because they can emit light long after being exposed to a light source.

❖ Dysprosium-doped crystals are used in certain types of lasers, particularly in infrared lasers for medical and industrial applications.

❖ Dysprosium is used in control rods in nuclear reactors to regulate the nuclear fission process by absorbing excess neutrons.

❖ Dysprosium compounds are used in some fluorescent and mercury-vapor lamps, improving their color output and light quality.

❖ Dysprosium has one of the highest magnetic strengths among the rare earth elements, making it crucial for miniaturized electronic components and high-performance magnets.

❖ Dysprosium is added to alloys with metals like stainless steel to enhance their corrosion resistance, especially in marine environments.

❖ Dysprosium is used in data storage technologies such as CDs/DVDs.

PHYSICAL PROPERTIES:

Physical Property	Description
Atomic Symbol	Dy
Atomic Number	66
Atomic Mass	162.50 amu
State at Room Temp	Solid (soft, metallic)
Color	Silvery-white
Odor	Odorless
Density	8.54 g/cm^3
Melting Point	1407°C
Boiling Point	2562°C
Solubility in Water	Insoluble
Conductivity	Good conductor of heat and electricity
Abundance in Earth	Found in minerals like monazite and bastnäsite, relatively abundant for a rare earth element

CHEMICAL PROPERTIES:

Chemical Property	Description
Reactivity	Dysprosium reacts slowly with air, forming a protective oxide layer (Dy_2O_3)
Oxidation States	+3 is the most stable oxidation state
Electronegativity	1.22 (Pauling scale)
Reaction with Water	Reacts slowly with water to form dysprosium hydroxide and hydrogen gas
Reaction with Acids	Dissolves in acids to form dysprosium salts
Reaction with Oxygen	Reacts with oxygen to form dysprosium oxide (Dy_2O_3)
Flammability	Flammable in powdered form
Toxicity	Considered mildly toxic, with no known biological role
Biological Importance	No known biological role
Isotopes	Dysprosium has seven stable isotopes, with Dysprosium-164 being the most abundant

HOLMIUM

Hi magnetic experts! I'm Holmium, with the highest magnetic strength of any element. I'm used in magnets and scientific instruments that need precision.

INTERESTING FACTS ABOUT HOLMIUM

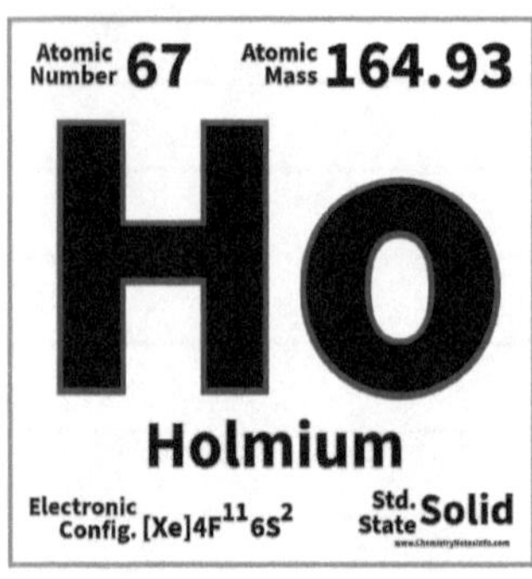

❖Holmium is named after Stockholm (Holmia in Latin), the capital of Sweden, in honour of the country where many rare earth elements were discovered.

❖Holmium was discovered in 1878 by Swedish chemist Per Teodor Cleve and Swiss chemist Marc Delafontaine, who isolated it from erbium oxide.

❖Holmium has the highest magnetic moment of any naturally occurring element, meaning it can become highly magnetised and is essential in powerful magnets.

❖ Due to its strong magnetic properties, Holmium is used in devices that generate strong magnetic fields, such as magnetic flux concentrators.

❖ Holmium-doped lasers are widely used in medical procedures, especially in laser surgeries for treating kidney stones, tumours, and other soft tissue operations.

❖ Holmium is used in nuclear reactors as part of control rods because of its ability to absorb neutrons effectively, helping control fission reactions.

❖ Holmium's paramagnetic properties make it useful in nuclear magnetic resonance (NMR) spectroscopy as a shift reagent, helping to analyse molecular structures.

❖ Unlike some other rare earth elements, Holmium is relatively stable in air and forms a protective oxide layer that prevents further corrosion.

❖ Holmium is found in the Earth's crust, typically in rare earth minerals like monazite and bastnäsite, though it is relatively rare.

❖ Despite being rare, holmium is more abundant in Earth's crust than silver, though it is dispersed and not found in concentrated deposits.

❖ Holmium-doped phosphors are used in some fluorescent lighting applications, contributing to their brightness and efficiency.

PHYSICAL PROPERTIES:

Physical Property	Description
Atomic Symbol	Ho
Atomic Number	67
Atomic Mass	164.93 amu
State at Room Temp	Solid (soft, metallic)
Color	Silvery-white
Odor	Odorless
Density	8.80 g/cm^3
Melting Point	1461°C
Boiling Point	2600°C
Solubility in Water	Insoluble
Conductivity	Good conductor of heat and electricity
Abundance in Earth	Found in minerals like monazite and bastnäsite, though relatively rare

CHEMICAL PROPERTIES:

Chemical Property	Description
Reactivity	Reacts slowly with air, forming a protective oxide layer (Ho_2O_3)
Oxidation States	+3 is the most common oxidation state
Electronegativity	1.23 (Pauling scale)
Reaction with Water	Reacts slowly with water to form holmium hydroxide and hydrogen gas
Reaction with Acids	Dissolves in acids to form holmium salts
Reaction with Oxygen	Reacts with oxygen to form holmium oxide (Ho_2O_3)
Flammability	Flammable in powdered form
Toxicity	Holmium compounds are considered mildly toxic if inhaled or ingested
Biological Importance	No known biological role
Isotopes	Holmium has only one naturally occurring isotope, Holmium-165

ERBIUM

Hello telecom enthusiasts! I'm Erbium, enhancing fibre optic cables for faster communication. I'm all about boosting the speed of modern data transfer.

INTERESTING FACTS ABOUT ERBIUM

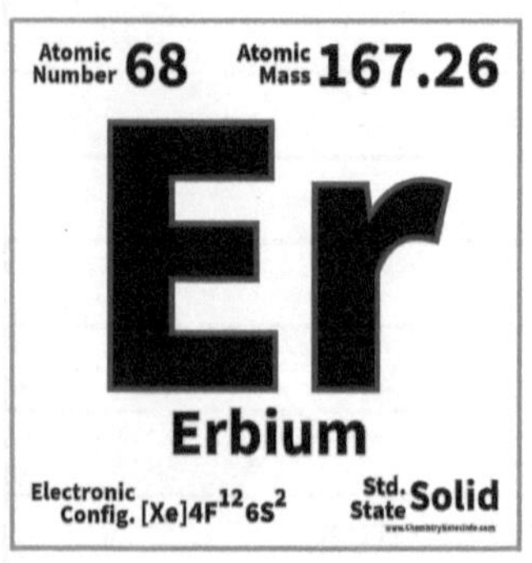

❖Like several other rare earth elements, Erbium is named after the village of Ytterby in Sweden, where the mineral gadolinite was discovered.

❖Erbium was discovered in 1843 by Swedish chemist Carl Gustaf Mosander, who isolated it from the mineral yttria while studying rare earth oxides.

❖Erbium salts, especially erbium chloride, have a distinctive pink colour and are often used to tint glass and ceramics for decorative purposes.

❖ One of Erbium's most important uses is in fibre-optic communication systems. Erbium-doped fibre amplifiers (EDFAs) are used to boost the signal strength in long-distance fiber-optic cables, enabling high-speed data transmission.

❖ Erbium-doped lasers are widely used in dermatology and dentistry. Erbium lasers are particularly effective in skin resurfacing treatments and dental procedures like cavity preparation and root canal surgery.

❖ Erbium compounds are used in phosphorescent materials to produce bright red and pink colours in screens, displays, and lighting.

❖ Erbium metal is relatively stable in air compared to other rare earth elements. It forms a protective oxide layer on its surface, which prevents it from rapid oxidation.

❖ Erbium is used in nuclear reactors as a neutron absorber. It helps to regulate the rate of the nuclear fission reaction.

❖ Erbium compounds are considered to be of low toxicity, though they must still be handled with care, especially in powdered form.

❖ Erbium has excellent neutron-absorbing properties, which makes it valuable in nuclear reactor control rods to manage fission reactions.

PHYSICAL PROPERTIES:

Physical Property	Description
Atomic Symbol	Er
Atomic Number	68
Atomic Mass	167.26 amu
State at Room Temp	Solid (soft, metallic)
Color	Silvery-white
Odor	Odorless
Density	9.07 g/cm^3
Melting Point	1529°C
Boiling Point	2868°C
Solubility in Water	Insoluble
Conductivity	Good conductor of heat and electricity
Abundance in Earth	Found in minerals like monazite and bastnäsite, relatively abundant for a rare earth element

CHEMICAL PROPERTIES:

Chemical Property	Description
Reactivity	Reacts slowly with oxygen, forming a protective oxide layer (Er_2O_3)
Oxidation States	+3 is the most common oxidation state
Electronegativity	1.24 (Pauling scale)
Reaction with Water	Reacts slowly with water to form erbium hydroxide and hydrogen gas
Reaction with Acids	Dissolves in acids to form erbium salts
Reaction with Oxygen	Reacts with oxygen to form erbium oxide (Er_2O_3)
Flammability	Flammable in powdered form
Toxicity	Erbium compounds are generally considered to be of low toxicity
Biological Importance	No known biological role
Isotopes	Erbium has six stable isotopes, with Erbium-166 being the most abundant

THULIUM

Hey innovators! I'm Thulium, rare and valuable in lasers and X-ray devices. I'm here to help push the boundaries of science and medicine.

INTERESTING FACTS ABOUT THULIUM

❖Thulium is named after "Thule," the mythical northern land thought to be the northernmost point of the ancient world, often associated with Scandinavia.

❖Thulium was discovered in 1879 by Swedish chemist Per Teodor Cleve while separating the elements in the rare earth oxide erbia.

❖Thulium is the least abundant of the lanthanides (rare earth elements), making it one of the rarest in the group, though still more abundant than other rare elements like platinum.

❖ Thulium is a soft, silvery metal that can be easily cut with a knife. It is also malleable and ductile, allowing it to be shaped into thin sheets or wires.

❖ Thulium-170, a radioactive isotope, is used in small, portable X-ray devices because it emits X-rays efficiently, making it useful in medical and industrial radiography.

❖ Thulium can be used as a neutron absorber in nuclear reactors due to its ability to capture neutrons, making it valuable in managing fission reactions.

❖ Thulium-doped lasers are used in certain surgical procedures, particularly in urology for treating conditions like kidney stones and benign prostatic hyperplasia (BPH).

❖ Thulium compounds are used to create bright green phosphors in certain types of lighting and displays, especially in cathode-ray tubes (CRTs) and specialised lamps.

❖ Like many rare earth elements, Thulium has magnetic properties and can conduct electricity well, though it is not typically used for these properties in practical applications.

❖ Thulium is relatively stable in air compared to other lanthanides, though it will oxidise slowly to form a protective oxide layer.

PHYSICAL PROPERTIES:

Physical Property	Description
Atomic Symbol	Tm
Atomic Number	69
Atomic Mass	168.93 amu
State at Room Temp	Solid (soft, metallic)
Color	Silvery-gray
Odor	Odorless
Density	9.32 g/cm^3
Melting Point	1545°C
Boiling Point	1950°C
Solubility in Water	Insoluble
Conductivity	Good conductor of heat and electricity
Abundance in Earth	Found in minerals like monazite and xenotime, one of the least abundant lanthanides

CHEMICAL PROPERTIES:

Chemical Property	Description
Reactivity	Thulium reacts slowly with air, forming a protective oxide layer (Tm_2O_3)
Oxidation States	+3 is the most common oxidation state
Electronegativity	1.25 (Pauling scale)
Reaction with Water	Reacts slowly with water to form thulium hydroxide and hydrogen gas
Reaction with Acids	Dissolves in acids to form thulium salts
Reaction with Oxygen	Reacts with oxygen to form thulium oxide (Tm_2O_3)
Flammability	Flammable in powdered form
Toxicity	Thulium compounds are considered mildly toxic if inhaled or ingested
Biological Importance	No known biological role
Isotopes	Thulium has one stable isotope, Thulium-169, and several radioactive isotopes

YTTERBIUM

Hello quantum scientists! I'm Ytterbium, used in atomic clocks and quantum computing research. I'm at the frontier of timekeeping and future tech.

INTERESTING FACTS ABOUT YTTERBIUM

❖Ytterbium, like several other rare earth elements, is named after the village of Ytterby in Sweden, where the mineral containing it was first discovered.

❖Ytterbium was discovered in 1878 by Swiss chemist Jean Charles Galissard de Marignac while analysing rare earth oxides.

❖Ytterbium plays a crucial role in highly accurate atomic clocks. Ytterbium-ion clocks are one of the most stable and precise timekeeping devices known.

❖ Ytterbium-doped materials are used in high-powered lasers for industrial cutting and welding applications. Ytterbium lasers are valued for their efficiency and power output.

❖ Ytterbium-169, a radioactive isotope, is used as a portable gamma-ray source for radiography, offering a safer and more portable alternative to traditional X-ray machines.

❖ Ytterbium is used in optical fibre amplifiers, enhancing the strength of signals in fibre optic cables, especially for long-distance telecommunications.

❖ Ytterbium is used as an alloying agent to improve the mechanical properties and strength of stainless steel and other metal alloys.

❖ Ytterbium isotopes are used in medical imaging and cancer treatments, where radioactive isotopes help target and destroy cancerous cells.

❖ Ytterbium is a soft, ductile, and malleable metal, meaning it can easily be shaped and deformed without breaking, making it useful in certain specialised applications.

❖ Unlike most lanthanides, which predominantly exhibit a +3 oxidation state, Ytterbium can exist in both +2 and +3 oxidation states, adding versatility to its chemical behaviour.

PHYSICAL PROPERTIES:

Physical Property	Description
Atomic Symbol	Yb
Atomic Number	70
Atomic Mass	173.04 amu
State at Room Temp	Solid (soft, metallic)
Color	Silvery-white
Odor	Odorless
Density	6.90 g/cm^3
Melting Point	824°C
Boiling Point	1196°C
Solubility in Water	Insoluble
Conductivity	Good conductor of heat and electricity
Abundance in Earth	Found in minerals like monazite and xenotime, relatively rare for a lanthanide

CHEMICAL PROPERTIES:

Chemical Property	Description
Reactivity	Ytterbium reacts slowly with air, forming a protective oxide layer (Yb_2O_3)
Oxidation States	+2 and +3 oxidation states are both common
Electronegativity	1.1 (Pauling scale)
Reaction with Water	Reacts slowly with water to form ytterbium hydroxide and hydrogen gas
Reaction with Acids	Dissolves in acids to form ytterbium salts
Reaction with Oxygen	Reacts with oxygen to form ytterbium oxide (Yb_2O_3)
Flammability	Flammable in powdered form
Toxicity	Ytterbium compounds are generally considered non-toxic, though safety precautions are necessary
Biological Importance	No known biological role
Isotopes	Ytterbium has seven stable isotopes, with Ytterbium-174 being the most abundant

LUTETIUM

Hello researchers! I'm Lutetium, a heavy hitter in PET scans and scientific research. From medical imaging to catalysts, I'm making breakthroughs happen.

INTERESTING FACTS ABOUT LUTETIUM

❖Lutetium is named after "Lutetia," the ancient Roman name for the city of Paris, in honour of the city where the element was first discovered.

❖Lutetium was independently discovered in 1907 by French chemist Georges Urbain and Austrian chemist Carl Auer von Welsbach while analyzing the rare earth element ytterbia.

❖Lutetium is one of the densest of the lanthanides, with a high melting point and high density compared to other rare earth elements.

❖ Lutetium is used in petroleum refining as a catalyst in processes like cracking, where it helps break down large hydrocarbons into smaller, more useful products.

❖ Lutetium-177 is used in targeted cancer therapy, especially for treating neuroendocrine tumours and prostate cancer. It delivers radiation directly to cancer cells with minimal impact on surrounding healthy tissue.

❖ Lutetium is very stable and resistant to corrosion. It forms a protective oxide layer when exposed to air, which helps maintain its lustre.

❖ Lutetium has the smallest atomic radius of all the lanthanides, which contributes to its unique chemical behaviour compared to other rare earth elements.

❖ Lutetium is harder and more brittle than most other rare earth metals, making it less malleable but more resistant to wear.

❖ Lutetium is one of the rarest of the lanthanides, found in low concentrations in minerals like monazite and bastnäsite.

❖ Lutetium is used in certain phosphor materials, which are crucial in lighting and display technologies, such as LED lights & TV screens.

PHYSICAL PROPERTIES:

Physical Property	Description
Atomic Symbol	Lu
Atomic Number	71
Atomic Mass	174.97 amu
State at Room Temp	Solid (hard, metallic)
Color	Silvery-white
Odor	Odorless
Density	9.84 g/cm³
Melting Point	1663°C
Boiling Point	3402°C
Solubility in Water	Insoluble
Conductivity	Good conductor of heat and electricity
Abundance in Earth	Found in low concentrations in minerals like monazite and bastnäsite

CHEMICAL PROPERTIES:

Chemical Property	Description
Reactivity	Lutetium reacts slowly with oxygen, forming a stable oxide layer (Lu_2O_3)
Oxidation States	+3 is the most common oxidation state
Electronegativity	1.27 (Pauling scale)
Reaction with Water	Reacts slowly with water to form lutetium hydroxide and hydrogen gas
Reaction with Acids	Dissolves in acids to form lutetium salts
Reaction with Oxygen	Reacts with oxygen to form lutetium oxide (Lu_2O_3)
Flammability	Flammable in powdered form
Toxicity	Lutetium compounds are mildly toxic, though it has no known biological role
Biological Importance	No known biological role
Isotopes	Lutetium has two naturally occurring isotopes: Lutetium-175 (stable) and Lutetium-176 (radioactive)

HAFNIUM

Hi nuclear engineers! I'm Hafnium, known for my ability to absorb neutrons. You'll find me in control rods of nuclear reactors and high-temperature alloys.

INTERESTING FACTS ABOUT HAFNIUM

❖Hafnium is named after "Hafnia," the Latin name for Copenhagen, where it was discovered in 1923 by Dirk Coster &George de Hevesy.

❖Hafnium was identified using X-ray spectroscopy, a technique that revealed its presence in zirconium ores, after being predicted by the periodic table.

❖Hafnium is almost always found in minerals containing zirconium, as the two elements have similar chemical properties and are difficult to separate.

❖ Hafnium is highly effective at absorbing neutrons, making it a valuable material in control rods for nuclear reactors to regulate fission reactions.

❖ Hafnium is highly resistant to corrosion, even at high temperatures, which makes it suitable for use in aerospace components and other harsh environments.

❖ Hafnium has a very high melting point of 2233°C, making it one of the refractory metals, materials capable of withstanding extreme heat without losing their strength.

❖ Due to its high melting point and resistance to heat and corrosion, hafnium is used in the nozzles of rockets and spacecraft engines.

❖ Hafnium-based compounds are used as gate insulators in modern high-performance computer chips, contributing to the miniaturisation of electronics.

❖ Although hafnium is more abundant than gold or silver, it is still considered rare and difficult to extract because it is chemically similar to zirconium.

❖ Hafnium is added to superalloys used in jet engines and gas turbines to improve the mechanical strength and durability of components exposed to high temperatures.

PHYSICAL PROPERTIES:

Physical Property	Description
Atomic Symbol	Hf
Atomic Number	72
Atomic Mass	178.49 amu
State at Room Temp	Solid (hard, metallic)
Color	Silvery-gray
Odor	Odorless
Density	13.31 g/cm^3
Melting Point	2233°C
Boiling Point	4600°C
Solubility in Water	Insoluble
Conductivity	Good conductor of heat and electricity
Abundance in Earth	Found in minerals like zircon and baddeleyite, often alongside zirconium

CHEMICAL PROPERTIES:

Chemical Property	Description
Reactivity	Hafnium reacts with oxygen and halogens at high temperatures, forming stable compounds
Oxidation States	+4 is the most common oxidation state
Electronegativity	1.3 (Pauling scale)
Reaction with Water	Reacts slowly with water at high temperatures
Reaction with Acids	Dissolves in hydrofluoric acid to form hafnium fluoride
Reaction with Oxygen	Reacts with oxygen to form hafnium dioxide (HfO_2), a highly stable compound
Flammability	Non-flammable in bulk form but can be pyrophoric (spontaneously combusting) in powdered form
Toxicity	Hafnium is generally non-toxic, but fine powders can be hazardous if inhaled
Biological Importance	No known biological role
Isotopes	Hafnium has six stable isotopes, with Hafnium-180 being the most abundant

TANTALUM

Hello tech lovers! I'm Tantalum, essential for making capacitors in your smartphones and electronics. I'm all about reliability and durability under pressure.

INTERESTING FACTS ABOUT TANTALUM

❖Tantalum is named after the mythological Greek figure Tantalus, because of the element's inability to absorb the acids, reminiscent of Tantalus's eternal punishment where he could not drink the water surrounding him.

❖Tantalum is almost immune to corrosion. It does not react with most acids, including hydrochloric and sulphuric acids, making it ideal for use in chemical equipment.

❖Tantalum was discovered in 1802 by Swedish chemist Anders Ekeberg, though it was initially confused with niobium due to their similar properties.

❖ Tantalum is a critical component in the manufacture of capacitors and high-performance resistors, which are essential for mobile phones, computers, and other electronic devices.

❖ Tantalum is used in superalloys that are employed in jet engines, turbines, and spacecraft due to its ability to maintain its strength at high temperatures.

❖ Tantalum is biocompatible, meaning it does not react with body tissues, making it useful in surgical implants like bone replacements, screws, and dental tools.

❖ Tantalum has one of the highest melting points of all elements, at around 3017°C, making it suitable for high-temperature applications like furnace parts.

❖ Tantalum pentoxide (Ta_2O_5) is used to make high-quality lenses for cameras and microscopes, as it is highly transparent and has a high refractive index.

❖ Tantalum and niobium are chemically very similar and are often found together in nature, making them difficult to separate. However, they have distinct industrial uses.

PHYSICAL PROPERTIES:

Physical Property	Description
Atomic Symbol	Ta
Atomic Number	73
Atomic Mass	180.95 amu
State at Room Temp	Solid (hard, metallic)
Color	Blue-gray, silvery-white
Odor	Odorless
Density	16.69 g/cm^3
Melting Point	3017°C
Boiling Point	5458°C
Solubility in Water	Insoluble
Conductivity	Good conductor of heat and electricity
Abundance in Earth	Found in minerals like coltan, tantalites, and euxenite

CHEMICAL PROPERTIES:

Chemical Property	Description
Reactivity	Highly resistant to corrosion; inert to nearly all acids except hydrofluoric acid
Oxidation States	+5 is the most common oxidation state, though +3 and +4 are also possible
Electronegativity	1.5 (Pauling scale)
Reaction with Water	Tantalum does not react with water
Reaction with Acids	Tantalum is resistant to almost all acids, except hydrofluoric acid
Reaction with Oxygen	Reacts with oxygen at high temperatures to form tantalum pentoxide (Ta_2O_5)
Flammability	Non-flammable in bulk form
Toxicity	Tantalum is generally non-toxic and safe for medical use
Biological Importance	No known biological role
Isotopes	Tantalum has two stable isotopes: Tantalum-181 (99.988%) and Tantalum-180m (extremely rare)

TUNGSTEN

Hi tough workers! I'm Tungsten, known for my incredible strength and high melting point. You'll find me in light bulbs, drills, and cutting tools—where the heat is on!

INTERESTING FACTS ABOUT TUNGSTEN

❖Tungsten has the highest melting point of any metal at 3422°C (6192°F), which makes it essential in high-temperature applications like light bulb filaments and rocket engine nozzles.

❖Tungsten was discovered by Spanish chemists Juan José and Fausto Elhuyar in 1783. They isolated the metal from the mineral wolframite.

❖The element is sometimes called "wolfram," which is derived from the mineral wolframite, its most important source. Its symbol "W" comes from this name.

❖ Tungsten is extremely dense, ranking just behind osmium and iridium, with a density of 19.25 g/cm³, making it ideal for use in heavy machinery and military applications like armour-piercing shells.

❖ Tungsten is incredibly hard and durable, which is why it's used in cutting tools, drills, and other high-strength applications.

❖ Tungsten carbide (a compound of tungsten and carbon) is one of the hardest materials known and is used extensively in industrial machinery, tools, and even jewellery due to its scratch-resistant properties.

❖ Tungsten's high melting point made it a key material for the filaments in incandescent light bulbs, as it can withstand the heat without melting or breaking.

❖ Tungsten is used to make super-strong alloys used in aerospace, automotive, and military industries, particularly in jet engines and missile components.

❖ China dominates tungsten production, supplying over 80% of the world's tungsten, making it a strategically important resource for industries worldwide.

❖ Tungsten is used in radiation shielding due to its density and high atomic number, which allow it to absorb and deflect harmful radiation.

PHYSICAL PROPERTIES:

Physical Property	Description
Atomic Symbol	W
Atomic Number	74
Atomic Mass	183.84 amu
State at Room Temp	Solid (hard, metallic)
Color	Steel-gray to white
Odor	Odorless
Density	19.25 g/cm^3
Melting Point	3422°C
Boiling Point	5555°C
Solubility in Water	Insoluble
Conductivity	Good conductor of heat and electricity
Abundance in Earth	Found in minerals like scheelite and wolframite

CHEMICAL PROPERTIES:

Chemical Property	Description
Reactivity	Tungsten is relatively inert, but it can react with halogens at high temperatures
Oxidation States	Common oxidation states are +6, +5, +4, and +3
Electronegativity	2.36 (Pauling scale)
Reaction with Water	Tungsten does not react with water
Reaction with Acids	Tungsten is resistant to most acids, but reacts slowly with hydrofluoric acid and nitric acid
Reaction with Oxygen	At high temperatures, tungsten reacts with oxygen to form tungsten oxide (WO_3)
Flammability	Non-flammable in bulk form but can be flammable in powder form
Toxicity	Generally considered non-toxic, though tungsten dust can be hazardous if inhaled
Biological Importance	No known biological role
Isotopes	Tungsten has five stable isotopes, with Tungsten-184 being the most abundant

RHENIUM

Hey jet-setters! I'm Rhenium, found in superalloys used in jet engines. I'm built to withstand extreme heat and keep planes flying high.

INTERESTING FACTS ABOUT RHENIUM

❖Rhenium was one of the last naturally occurring elements to be discovered. It was identified by German chemists Ida and Walter Noddack and Otto Berg in 1925.

❖The element was named after the Rhine River in Europe (Latin: Rhenus) to honour the region where it was discovered.

❖Rhenium is one of the rarest elements in the Earth's crust, with an average concentration of about 1 part per billion. It is typically obtained as a byproduct of molybdenum and copper ore refining.

❖ Rhenium has one of the highest melting points of any element, at 3186°C, making it valuable in high-temperature superalloys for jet engines and gas turbines.

❖ Rhenium is added to nickel-based superalloys to improve the performance of jet engines. About 70% of the rhenium produced annually is used for this purpose.

❖ Rhenium is used as a catalyst in the production of lead-free gasoline and in the petrochemical industry to refine crude oil.

❖ Despite being very hard, rhenium is ductile and can be bent and shaped without breaking, unlike many other metals with similar hardness.

❖ Rhenium has high electrical resistance, making it useful in filaments for mass spectrometers, X-ray machines, and other electronic devices.

❖ The isotope Rhenium-187 has an incredibly long half-life of 41.6 billion years, making it stable on a cosmic timescale.

❖ NASA has studied rhenium-based alloys for space propulsion systems because of its ability to withstand extreme heat.

❖ Rhenium is alloyed with platinum or tungsten to make thermocouples, which are used to measure high temperatures in furnaces and jet engines.

PHYSICAL PROPERTIES:

Physical Property	Description
Atomic Symbol	Re
Atomic Number	75
Atomic Mass	186.21 amu
State at Room Temp	Solid (metallic)
Color	Silvery-gray
Odor	Odorless
Density	21.02 g/cm³
Melting Point	3186°C
Boiling Point	5630°C
Solubility in Water	Insoluble
Conductivity	Good conductor of heat and electricity
Abundance in Earth	Found in very low concentrations, typically as a byproduct of molybdenum and copper ores

CHEMICAL PROPERTIES:

Chemical Property	Description
Reactivity	Rhenium is relatively inert but reacts with oxygen and halogens at high temperatures
Oxidation States	Common oxidation states are +7, +6, +4, and +3
Electronegativity	1.9 (Pauling scale)
Reaction with Water	Rhenium does not react with water
Reaction with Acids	Dissolves in nitric acid, sulfuric acid, and aqua regia
Reaction with Oxygen	Forms rhenium oxide (Re_2O_7) when heated in oxygen
Flammability	Non-flammable in bulk form
Toxicity	Rhenium compounds can be toxic if ingested, but elemental rhenium is generally non-toxic
Biological Importance	No known biological role
Isotopes	Rhenium has two naturally occurring isotopes: Rhenium-185 (stable) and Rhenium-187 (radioactive)

OSMIUM

Hello weight lifters! I'm Osmium, the densest naturally occurring element. I'm found in alloys that need to endure incredible pressure and stress.

INTERESTING FACTS ABOUT OSMIUM

❖Osmium is the densest naturally occurring element, with a density of 22.59 g/cm³. It is even denser than lead and gold.

❖Osmium was discovered in 1803 by British chemist Smithson Tennant while he was studying the residues left behind after platinum ore was dissolved in aqua regia.

❖The name "osmium" is derived from the Greek word "osme," meaning smell, because of the strong, unpleasant odour of its volatile oxide, osmium tetroxide (OsO_4).

❖ Osmium was once used in alloys with iridium to make fountain pen nibs due to its extreme hardness and resistance to wear.

❖ Osmium tetroxide is used in electron microscopy as a stain for biological tissues because it bind to fats & enhances contrast in cells

❖ Osmium is used as a catalyst in a number of industrial chemical reactions, including the production of ammonia and the hydrogenation of organic compounds.

❖ Osmium is one of the hardest metals known, making it valuable in the production of durable alloys for things like electrical contacts and surgical implants.

❖ Osmium is the rarest of the platinum-group metals, making it one of the most valuable and sought-after metals for specialised applications.

❖ Osmium has a melting point of 3033°C, making it suitable for high-temperature applications, though it is less commonly used due to its toxicity in some forms.

❖ Osmium is highly resistant to corrosion, even at high temperatures, and is used in alloys to provide durability in harsh environments.

❖ Osmium is one of the most expensive elements, with prices often surpassing that of gold due to its rarity and specialised uses.

PHYSICAL PROPERTIES:

Physical Property	Description
Atomic Symbol	Os
Atomic Number	76
Atomic Mass	190.23 amu
State at Room Temp	Solid (hard, metallic)
Color	Bluish-white, silvery
Odor	Odorless (elemental osmium); Osmium tetroxide has a sharp, chlorine-like odor
Density	22.59 g/cm³
Melting Point	3033°C
Boiling Point	5012°C
Solubility in Water	Insoluble
Conductivity	Good conductor of heat and electricity
Abundance in Earth	Extremely rare, found in platinum ores and as a byproduct of nickel refining

CHEMICAL PROPERTIES:

Chemical Property	Description
Reactivity	Relatively inert at room temperature but reacts with oxygen when heated to form osmium tetroxide
Oxidation States	Common oxidation states are +4, +6, and +8
Electronegativity	2.2 (Pauling scale)
Reaction with Water	Does not react with water
Reaction with Acids	Resistant to most acids; can react with concentrated sulfuric and nitric acid at high temperatures
Reaction with Oxygen	Reacts with oxygen at high temperatures to form toxic osmium tetroxide (OsO_4)
Flammability	Non-flammable in bulk form but osmium tetroxide is volatile and flammable
Toxicity	Osmium metal is non-toxic, but osmium tetroxide is highly toxic and corrosive
Biological Importance	No known biological role
Isotopes	Osmium has seven naturally occurring isotopes, with Osmium-192 being the most abundant

IRIDIUM

Hi explorers! I'm Iridium, one of the most corrosion-resistant elements. From deep-sea cables to space exploration, I'm protecting technology in the harshest environments.

INTERESTING FACTS ABOUT IRIDIUM

❖Iridium is the second densest element after osmium, with a density of 22.56 g/cm³. This makes it extremely durable and resistant to deformation.

❖Iridium was discovered in 1803 by Smithson Tennant, along with osmium, in the residue left after dissolving platinum ore in aqua regia.

❖Iridium's name comes from the Greek word "iris," meaning rainbow, because its compounds exhibit a wide range of colours.

❖ Iridium is often used in alloys for making durable fountain pen tips, though modern pen tips use a variety of iridium group metals.

❖ With a melting point of 2446°C, iridium is used in high-temperature applications such as crucibles for growing crystals or in spark plugs.

❖ A global layer of iridium-rich clay from around 66 million years ago is linked to the asteroid impact that is thought to have caused the extinction of the dinosaurs. The element is more abundant in meteorites than in Earth's crust.

❖ Iridium is often alloyed with platinum to create super-strong and corrosion-resistant materials used in electrical contacts, medical devices, and other high-stress applications.

❖ Iridium is used in certain types of radiation shielding and radioactive isotopes like Iridium-192 are used in cancer treatment for brachytherapy.

❖ Despite being incredibly hard, iridium is also brittle, making it difficult to work with unless alloyed with other metals.

❖ Iridium is used in the construction of spacecraft and satellites, particularly in components that need to endure extreme temperatures and radiation in space.

PHYSICAL PROPERTIES:

Physical Property	Description
Atomic Symbol	Ir
Atomic Number	77
Atomic Mass	192.22 amu
State at Room Temp	Solid (metallic)
Color	Silvery-white
Odor	Odorless
Density	22.56 g/cm³
Melting Point	2446°C
Boiling Point	4428°C
Solubility in Water	Insoluble
Conductivity	Good conductor of heat and electricity
Abundance in Earth	Extremely rare, found in platinum ores and meteorites

CHEMICAL PROPERTIES:

Chemical Property	Description
Reactivity	Iridium is chemically inert at room temperature and resists most acids, including aqua regia
Oxidation States	Common oxidation states are +3, +4, +6, and +1
Electronegativity	2.2 (Pauling scale)
Reaction with Water	Does not react with water
Reaction with Acids	Resistant to most acids; can slowly dissolve in hot aqua regia
Reaction with Oxygen	Reacts with oxygen at very high temperatures to form iridium dioxide (IrO_2)
Flammability	Non-flammable in bulk form
Toxicity	Elemental iridium is not toxic, but some iridium compounds may be harmful if ingested or inhaled
Biological Importance	No known biological role
Isotopes	Iridium has two stable isotopes: Iridium-191 and Iridium-193, with Iridium-193 being the most abundant

PLATINUM

Hello luxury seekers! I'm Platinum, prized for my rarity and beauty in jewellery, but I also clean up car emissions in catalytic converters. I'm both precious and practical.

INTERESTING FACTS ABOUT PLATINUM

❖Platinum is considered one of the most precious metals due to its rarity and wide range of industrial, medical, and jewellery applications.

❖Platinum was first used by Pre-Columbian South American cultures before it was rediscovered by European scientists in the 16th century. It wasn't officially recognised in Europe until the mid-18th century.

❖Platinum is highly resistant to corrosion and tarnishing, even at high temperatures, which makes it extremely valuable in industrial applications.

❖ Around 50% of the platinum produced today is used in catalytic converters for cars, which help reduce harmful emissions by converting them into less toxic substances.

❖ Platinum is a popular metal for high-end jewellery, especially engagement rings and wedding bands. Its rarity makes it more valuable than gold in certain markets.

❖ Platinum compounds, like cisplatin, are used in chemotherapy treatments for various types of cancer, due to their ability to inhibit cancer cell division.

❖ Due to its high melting point and resistance to corrosion, platinum is used in spacecraft components and satellites that need to endure extreme conditions in space.

❖ Some countries mint platinum coins as a form of investment. The US, for example, produces platinum American Eagle coins.

❖ The term "platinum" is often used to symbolise the highest level of achievement or luxury, as seen in "platinum records" in music industry.

❖ Platinum is alloyed with other metals like iridium or ruthenium to improve hardness and durability, particularly in industrial settings.

PHYSICAL PROPERTIES:

Physical Property	Description
Atomic Symbol	Pt
Atomic Number	78
Atomic Mass	195.08 amu
State at Room Temp	Solid (metallic)
Color	Silvery-white, lustrous
Odor	Odourless
Density	21.45 g/cm^3
Melting Point	1768°C
Boiling Point	3825°C
Solubility in Water	Insoluble
Conductivity	Excellent conductor of electricity and heat
Abundance in Earth	Rare, often found in nickel and copper ores

CHEMICAL PROPERTIES:

Chemical Property	Description
Reactivity	Platinum is chemically inert and does not react with most substances
Oxidation States	Common oxidation states are +2 and +4
Electronegativity	2.28 (Pauling scale)
Reaction with Water	Does not react with water
Reaction with Acids	Resistant to most acids, but dissolves slowly in aqua regia (a mixture of nitric and hydrochloric acid)
Reaction with Oxygen	Platinum does not readily oxidize in air, even at high temperatures
Flammability	Non-flammable in bulk form
Toxicity	Elemental platinum is generally non-toxic; some platinum compounds, however, can be toxic
Biological Importance	No known biological role, though platinum-based drugs are used in cancer treatments
Isotopes	Platinum has six naturally occurring isotopes, with Platinum-195 being the most abundant

GOLD

Hey treasure hunters! I'm Gold, the symbol of wealth and beauty. From ancient coins to modern electronics, I'm as conductive as I am glamorous.

INTERESTING FACTS ABOUT GOLD

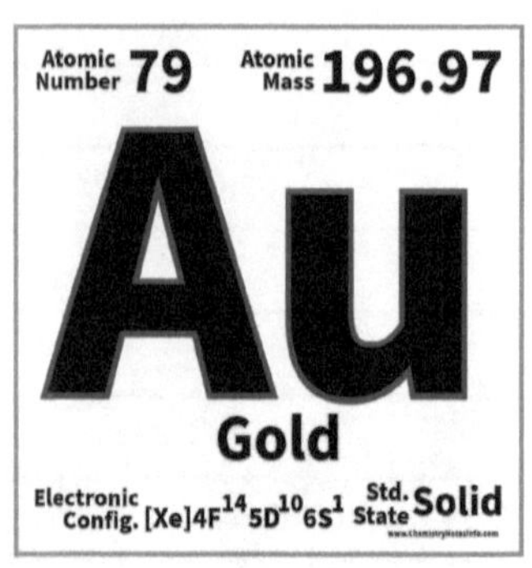

❖Gold has been a symbol of wealth and power for thousands of years, with civilisations across the world valuing it for its rarity, beauty, and workability.

❖Gold is most malleable of all metals. **1gm** of gold can be hammered into a sheet of **1m²** in size without breaking.

❖Gold is also incredibly ductile; it can be drawn into a wire as thin as a human hair without breaking.

❖Due to its excellent electrical conductivity and resistance to tarnish, gold is widely used in electronics, including connectors, switches, and relays in computers, smartphones, and other devices.

❖ Unlike many other metals, gold does not corrode, rust, or tarnish. It remains bright and shiny even after thousands of years, making it ideal for jewellery and art.

❖ For much of modern history, the value of many currencies was based on the gold standard, where currency was directly tied to a fixed amount of gold.

❖ Gold is non-toxic and can be consumed in small quantities. Edible gold leaf is used in luxury desserts, drinks, and decorations.

❖ Gold compounds have been used in the treatment of certain medical conditions, such as rheumatoid arthritis, due to their anti-inflammatory properties.

❖ Although present in trace amounts, vast quantities of gold are dissolved in seawater. However, extracting gold from seawater is not economically feasible with current technology.

❖ Gold coins have been used as currency for over 2,500 years, with some of the earliest known coins originating from Lydia, an ancient kingdom in what is now Turkey.

❖ Central banks and governments hold large reserves of gold bullion.

PHYSICAL PROPERTIES:

Physical Property	Description
Atomic Symbol	Au (from the Latin "Aurum")
Atomic Number	79
Atomic Mass	196.97 amu
State at Room Temp	Solid (metallic)
Color	Yellow, shiny
Odor	Odorless
Density	19.32 g/cm³
Melting Point	1064°C
Boiling Point	2856°C
Solubility in Water	Insoluble
Conductivity	Excellent conductor of electricity and heat
Abundance in Earth	Rare, found in veins in rocks and alluvial deposits

CHEMICAL PROPERTIES:

Chemical Property	Description
Reactivity	Gold is one of the least reactive chemical elements
Oxidation States	Common oxidation states are +1 and +3
Electronegativity	2.54 (Pauling scale)
Reaction with Water	Does not react with water
Reaction with Acids	Resistant to most acids, but dissolves in aqua regia (a mixture of nitric and hydrochloric acid)
Reaction with Oxygen	Does not tarnish or oxidize in air
Flammability	Non-flammable in bulk form
Toxicity	Elemental gold is non-toxic, though some gold compounds can be hazardous
Biological Importance	No known biological role
Isotopes	Gold has only one stable isotope: Gold-197

MERCURY

Hello liquid metal fans! I'm Mercury, the only metal that's liquid at room temperature. I've been used in thermometers and barometers, but handle with care—I can be toxic!

INTERESTING FACTS ABOUT MERCURY

❖Mercury is unique as the only metal that is liquid at room temperature (around 20°C). This distinctive property has earned it the nickname "quicksilver."

❖Chemical symbol "Hg" comes from the Latin word "hydrargyrum," meaning "water-silver," due to its liquid, silvery appearance.

❖Mercury is highly toxic to humans and animals. Exposure can lead to mercury poisoning, which affects the brain, nervous system, and kidneys.

❖ Due to its liquid form and thermal expansion properties, mercury was historically used in thermometers, barometers, and other pressure-sensing devices, though many of these have been phased out for safety reasons.

❖ Mercury was historically obtained from cinnabar (mercury sulphide ore). Large deposits of cinnabar were mined in places like Spain and Italy for thousands of years.

❖ Mercury has been used for over 3,500 years. Ancient Egyptians used mercury in cosmetics, and the Chinese believed it had healing properties, even though it's toxic.

❖ Mercury has been used in dental amalgams for fillings due to its ability to form stable alloys with other metals. However, its use is declining due to concerns over toxicity.

❖ Mercury is used in small-scale gold mining to form an amalgam with gold, which is then heated to evaporate the mercury, leaving behind pure gold. This process, however, poses severe environmental risks.

❖ Mercury is a good conductor of electricity and is used in electrical switches, relays, and rectifiers.

❖ Mercury easily forms alloys with other metals, called amalgams.

PHYSICAL PROPERTIES:

Physical Property	Description
Atomic Symbol	Hg (from the Latin "hydrargyrum")
Atomic Number	80
Atomic Mass	200.59 amu
State at Room Temp	Liquid
Color	Silvery-white, metallic
Odor	Odorless
Density	13.53 g/cm³
Melting Point	$-38.83°C$
Boiling Point	356.73°C
Solubility in Water	Insoluble
Conductivity	Good conductor of electricity but poor conductor of heat
Abundance in Earth	Relatively rare, found primarily in cinnabar (HgS)

CHEMICAL PROPERTIES:

Chemical Property	Description
Reactivity	Mercury is relatively inert in pure form, but its compounds are quite reactive
Oxidation States	Common oxidation states are $+1$ and $+2$
Electronegativity	2.00 (Pauling scale)
Reaction with Water	Does not react with water
Reaction with Acids	Mercury dissolves in nitric acid, forming mercury nitrate, but does not dissolve in hydrochloric acid
Reaction with Oxygen	Reacts with oxygen at high temperatures to form mercury(II) oxide (HgO)
Flammability	Non-flammable in liquid form
Toxicity	Highly toxic to humans and animals
Biological Importance	No known biological role, and exposure can cause severe health effects
Isotopes	Mercury has several isotopes, with Mercury-202 being the most abundant

THALLIUM

Greetings chemists! I'm Thallium, once used in rat poisons and now in medical imaging. I've got a dangerous side but can be quite useful in the right hands.

INTERESTING FACTS ABOUT THALLIUM

❖Thallium was discovered in 1861 by William Crookes and Claude-Auguste Lamy using flame spectroscopy. Its bright green spectral line inspired its name, which comes from the Greek word "thallos," meaning "green shoot."

❖Thallium is extremely toxic to humans, with effects similar to heavy metal poisoning. It has been used in poisons and was notoriously employed in several criminal cases.

❖Thallium was historically used in rat poisons and insecticides, but its use has been largely discontinued due to its extreme toxicity to humans and other animals.

❖ Thallium's compounds are odourless and tasteless, making them particularly dangerous as they are difficult to detect in food or water.

❖ Thallium is a soft metal that can be cut with a knife. It tarnishes quickly in air, forming a grey oxide layer.

❖ Thallium is used in specialised optics, such as infrared lenses, where its compounds like thallium bromoiodide (KRS-5) are prized for their optical properties.

❖ Thallium is sometimes called the "poisoner's poison" because of its past use in criminal cases. Despite its deadly nature, its medical isotope Thallium-201 has saved countless lives by detecting heart blockages.

❖ Thallium has been used in thermoelectric devices due to its ability to convert temperature differences into electrical voltage.

❖ Thallium-201 is a radioactive isotope used in nuclear medicine, particularly in stress tests of heart to diagnose coronary artery disease.

❖ Thallium poisoning can cause hair loss (alopecia), as well as damage to the skin and nails, which are often early signs of exposure.

❖ Thallium has chemical properties that resemble those of lead and is often found in minerals alongside lead and zinc.

PHYSICAL PROPERTIES:

Physical Property	Description
Atomic Symbol	Tl
Atomic Number	81
Atomic Mass	204.38 amu
State at Room Temp	Solid (metallic)
Color	Bluish-gray, similar to lead
Odor	Odorless
Density	11.85 g/cm³
Melting Point	304°C
Boiling Point	1473°C
Solubility in Water	Insoluble as a pure metal, but soluble as thallium compounds (e.g., thallium sulfate)
Conductivity	Poor conductor of electricity compared to other metals
Abundance in Earth	Relatively rare, often extracted as a byproduct of lead, zinc, and copper mining

CHEMICAL PROPERTIES:

Chemical Property	Description
Reactivity	Thallium is highly reactive, forming oxides when exposed to air
Oxidation States	Common oxidation states are +1 (thallous) and +3 (thallic)
Electronegativity	1.62 (Pauling scale)
Reaction with Water	Does not react with water in its metallic state, but forms hydroxides when dissolved as a compound
Reaction with Acids	Dissolves in most acids, forming thallium salts
Reaction with Oxygen	Quickly oxidizes in air to form a thin gray oxide layer (thallium oxide, Tl_2O)
Flammability	Non-flammable in metallic form, though certain thallium compounds may be hazardous
Toxicity	Highly toxic to humans and animals
Biological Importance	No known biological role, but toxic exposure can cause serious health issues
Isotopes	Thallium has two stable isotopes: Thallium-203 and Thallium-205, and several radioactive isotopes

LEAD

Hello builders! I'm Lead, known for my heaviness and shielding abilities. From pipes to radiation protection, I've been used throughout history—but always handle me with care.

INTERESTING FACTS ABOUT LEAD

❖Lead has been used by humans for over 6,000 years. The Romans used it extensively in plumbing, leading to the origin of the word "plumbing" from the Latin word for lead, "plumbum."

❖Lead is highly toxic, especially to the nervous system. Chronic exposure can cause lead poisoning, which affects multiple organs and is particularly dangerous for children.

❖Lead is a soft, malleable metal that can be easily shaped and deformed. It can be scratched with a fingernail.

❖ Lead is widely used in lead-acid batteries, which power cars, trucks, and other vehicles. These batteries are rechargeable and are also used in backup power systems.

❖ Lead is one of the heaviest common metals, with a density of 11.34 g/cm³. This makes it useful for applications like radiation shielding and counterweights.

❖ Lead has a relatively low melting point of 327.5°C, which makes it easy to melt and cast into shapes.

❖ For many years, lead was used in household paints due to its durability and bright colours. However, its use has been largely banned because of the health risks, especially for children who ingest lead-contaminated dust.

❖ Lead is highly effective at blocking radiation, particularly X-rays and gamma rays. It is used in protective shields in medical imaging and nuclear reactors.

❖ In ancient Rome, lead was used extensively in water pipes. It is still present in few older water systems which poses a public health risk.

❖ Lead is often alloyed with other metals, such as tin, to make solder, which is used to join electrical components. Lead alloys are also used in bullets and shot.

PHYSICAL PROPERTIES:

Physical Property	Description
Atomic Symbol	Pb (from the Latin "plumbum")
Atomic Number	82
Atomic Mass	207.2 amu
State at Room Temp	Solid (metallic)
Color	Bluish-gray, but tarnishes to a dull grey when exposed to air
Odor	Odourless
Density	11.34 g/cm³
Melting Point	327.5°C
Boiling Point	1749°C
Solubility in Water	Insoluble
Conductivity	Poor conductor of electricity compared to other metals
Abundance in Earth	Relatively common, often found in ores such as galena (PbS)

CHEMICAL PROPERTIES:

Chemical Property	Description
Reactivity	Lead reacts slowly with oxygen, forming lead oxide (PbO) on its surface
Oxidation States	Common oxidation states are +2 (lead(II)) and +4 (lead(IV))
Electronegativity	1.87 (Pauling scale)
Reaction with Water	Lead does not react with pure water but can be corroded by acidic water
Reaction with Acids	Reacts with most acids, forming lead salts, but is resistant to sulfuric acid
Reaction with Oxygen	Slowly reacts with oxygen in air to form a protective oxide layer
Flammability	Non-flammable in metallic form
Toxicity	Extremely toxic, particularly to the nervous system, kidneys, and reproductive organs
Biological Importance	No known biological role, and lead exposure can cause severe health problems

BISMUTH

Hi colourful crystals! I'm Bismuth, non-toxic and known for my rainbow-hued crystals. I'm found in medicines and as a lead replacement in safety applications.

INTERESTING FACTS ABOUT BISMUTH

❖Bismuth is often considered the heaviest non-radioactive element. However, recent studies have shown that it is weakly radioactive, its half-life is so long (1.9×10^{19} years) that it is effectively stable.

❖Bismuth is much less toxic than its neighbours in the periodic table, such as lead, and is considered one of the safest heavy metals for human use. It is often used in medicines.

❖Bismuth forms beautiful, geometric crystals with an iridescent rainbow-like surface when it oxidises. These crystals are highly sought after by mineral collectors.

❖ Bismuth is unusual in that it expands when it solidifies, similar to water. This makes it useful in casting and mould-making, as it fills moulds more completely than metals that shrink upon cooling.

❖ Bismuth compounds, such as bismuth subsalicylate, are used in over-the-counter medications like Pepto-Bismol to treat stomach ulcers, indigestion, and diarrhoea.

❖ Bismuth oxychloride is a common ingredient in cosmetics, such as eyeshadows and face powders, where it adds a shimmering, pearlescent effect.

❖ Bismuth is increasingly used as a non-toxic alternative to lead in various applications, such as in bullets, fishing weights, and solder.

❖ Bismuth has the lowest thermal conductivity of any metal. This property is useful in certain electronic and thermoelectric applications, where heat must be managed effectively.

❖ Bismuth is one of the most diamagnetic substances, meaning it is repelled by a magnetic field. This property makes it useful in scientific instruments.

❖ Some isotopes are used in nuclear medicine, like in radiotherapy for cancer treatments due to their radioactive decay properties.

PHYSICAL PROPERTIES:

Physical Property	Description
Atomic Symbol	Bi
Atomic Number	83
Atomic Mass	208.98 amu
State at Room Temp	Solid (metallic)
Color	Silvery-white with a pinkish tinge when freshly cut
Odor	Odorless
Density	9.78 g/cm³
Melting Point	271.4°C
Boiling Point	1564°C
Solubility in Water	Insoluble as a metal; some compounds are soluble
Conductivity	Poor conductor of heat and electricity compared to other metals
Abundance in Earth	Relatively rare, found in ores like bismuthinite (Bi_2S_3) and as a byproduct of lead refining

CHEMICAL PROPERTIES:

Chemical Property	Description
Reactivity	Bismuth is relatively stable and does not react with oxygen or water under normal conditions
Oxidation States	Common oxidation states are +3 and +5
Electronegativity	2.02 (Pauling scale)
Reaction with Water	Bismuth does not react with water but reacts with acids
Reaction with Acids	Dissolves slowly in concentrated acids, forming bismuth salts
Reaction with Oxygen	Forms a thin oxide layer when exposed to air, giving it a pinkish hue
Flammability	Non-flammable in metallic form
Toxicity	Low toxicity compared to other heavy metals, but some compounds can be toxic
Biological Importance	No known biological role, though bismuth compounds are used medicinally
Isotopes	Bismuth has only one naturally occurring isotope, Bismuth-209, which is weakly radioactive

POLONIUM

Hey radiation experts! I'm Polonium, a rare and highly radioactive element. I'm used in anti-static devices and can release intense energy—handle with extreme care!

INTERESTING FACTS ABOUT POLONIUM

❖Polonium was discovered by Marie and Pierre Curie in 1898. It was named after Marie Curie's homeland, Poland ("Polonia" in Latin), which was under foreign occupation at the time.

❖Polonium is one of the most radioactive elements. Polonium-210, its most common isotope, emits alpha particles and is extremely dangerous if ingested or inhaled.

❖Polonium-210 gained international attention when it was used to assassinate former Russian spy Alexander Litvinenko in 2006, highlighting its deadly toxicity.

❖ Polonium is extremely rare in nature. It is found in minute quantities in uranium ores but is usually produced artificially in nuclear reactors by bombarding bismuth with neutrons.

❖ A small amount of Polonium-210 can generate significant heat due to its intense radioactive decay. This makes it useful as a heat source in space missions.

❖ Polonium-210 decays by alpha emission with a half-life of 138 days, turning into lead-206, which is stable.

❖ While polonium's alpha radiation cannot penetrate the skin, ingestion or inhalation of polonium is lethal, as the alpha particles cause severe damage to internal tissues.

❖ During the Cold War, polonium was used in the initiators of early nuclear weapons, where it helped trigger the chain reaction in atomic bombs.

❖ Polonium was once used in devices designed to eliminate static electricity in industrial environments, though safer alternatives have now replaced it.

❖ Early on, polonium was extracted in tiny amounts from uranium ores such as pitchblende, but modern production occurs in nuclear reactors.

PHYSICAL PROPERTIES:

Physical Property	Description
Atomic Symbol	Po
Atomic Number	84
Atomic Mass	209 amu (for Polonium-209, the most stable isotope)
State at Room Temp	Solid (metallic)
Color	Silvery-grey, though rarely seen in pure form due to its intense radioactivity
Odor	Odourless
Density	9.32 g/cm^3
Melting Point	254°C
Boiling Point	962°C
Solubility in Water	Insoluble
Conductivity	Poor conductor of electricity and heat compared to other metals
Abundance in Earth	Extremely rare, primarily found in uranium ores and produced artificially in reactors

CHEMICAL PROPERTIES:

Chemical Property	Description
Reactivity	Polonium is chemically similar to tellurium and bismuth and is fairly reactive
Oxidation States	Common oxidation states are +2, +4, and +6
Electronegativity	2.0 (Pauling scale)
Reaction with Water	Does not react with water, but forms polonium hydroxide when exposed to moist air
Reaction with Acids	Dissolves in dilute acids, forming polonium salts
Reaction with Oxygen	Oxidises in air to form polonium dioxide (PoO_2)
Flammability	Non-flammable, though polonium is extremely hazardous due to its radioactivity
Toxicity	Extremely toxic, primarily due to its intense radioactivity rather than chemical toxicity
Biological Importance	No known biological role, and exposure is highly dangerous
Isotopes	Polonium has over 30 isotopes, but Polonium-210 is the most common and radioactive

ASTATINE

Hello researchers! I'm Astatine, one of the rarest naturally occurring elements. My radioactivity makes me a focus of cancer treatment research, despite my elusive nature.

INTERESTING FACTS ABOUT ASTATINE

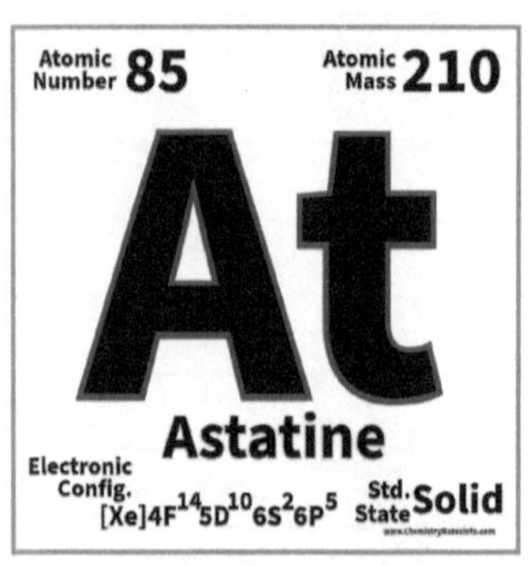

❖Astatine is the rarest naturally occurring element in the Earth's crust. It is estimated that less than 1 gram exists naturally at any given time on the planet.

❖Astatine is part of the halogen group (Group 17) in the periodic table, alongside fluorine, chlorine, bromine, and iodine, but it is the only halogen that is radioactive.

❖All isotopes of astatine are highly radioactive, with short half-lives. The most stable isotope, Astatine-210, has a half-life of only about 8.1 hours.

❖ The name "Astatine" comes from the Greek word "astatos," meaning "unstable," which reflects its radioactive instability.

❖ Astatine has potential medical uses in cancer treatment. Astatine-211 is being studied for targeted alpha-particle therapy to destroy cancer cells without damaging surrounding healthy tissues.

❖ Astatine is one of the least studied elements due to its extreme rarity and radioactivity, making it difficult to handle and analyse.

❖ Chemically, astatine behaves similarly to iodine, which suggests it could accumulate in thyroid gland if introduced into human body.

❖ Astatine is primarily produced artificially in particle accelerators by bombarding bismuth-209 with alpha particles to create Astatine-210.

❖ Astatine is the heaviest halogen, though its properties are not fully understood due to its scarcity and radioactivity.

❖ Astatine decays into radon (a radioactive noble gas), further contributing to its radioactivity.

❖ Like other halogens, astatine is expected to be highly toxic, but its toxicity is compounded by its radioactivity.

❖ Astatine is primarily used for research purposes, particularly in nuclear physics and radio-pharmaceuticals.

PHYSICAL PROPERTIES:

Physical Property	Description
Atomic Symbol	At
Atomic Number	85
Atomic Mass	210 amu (for Astatine-210, the most stable isotope)
State at Room Temp	Solid (assumed, though its exact appearance is unclear due to its rarity and radioactivity)
Color	Likely dark or metallic, though its exact color has not been observed directly
Odor	Unknown due to its extreme rarity and radioactivity
Density	Estimated to be around 7 g/cm³, though this is not confirmed
Melting Point	Estimated to be 302°C
Boiling Point	Estimated to be 337°C
Solubility in Water	Likely behaves similarly to iodine and is slightly soluble in water
Conductivity	Likely a poor conductor, similar to other halogens
Abundance in Earth	Extremely rare, with natural occurrence estimated at less than 1 gram globally

CHEMICAL PROPERTIES:

Chemical Property	Description
Reactivity	Astatine is expected to be less reactive than iodine but still reactive due to being a halogen
Oxidation States	Common oxidation states are -1, +1, +3, +5, &+7
Electronegativity	Estimated at 2.2 (Pauling scale)
Reaction with Water	Likely forms hydrides such as astatine hydride
Reaction with Acids	Likely forms halide salts, similar to iodine
Reaction with Oxygen	Expected to form astatine oxides, though these are difficult to study
Flammability	Non-flammable, but radioactive decay poses a health hazard
Toxicity	Expected to be extremely toxic, both chemically and due to its radioactivity
Biological Importance	No known biological role; highly dangerous due to its radioactivity

RADON

Hi homeowners! I'm Radon, a radioactive gas that can sneak into basements. I'm part of the natural decay chain of uranium, so keep an eye on indoor air quality!

INTERESTING FACTS ABOUT RADON

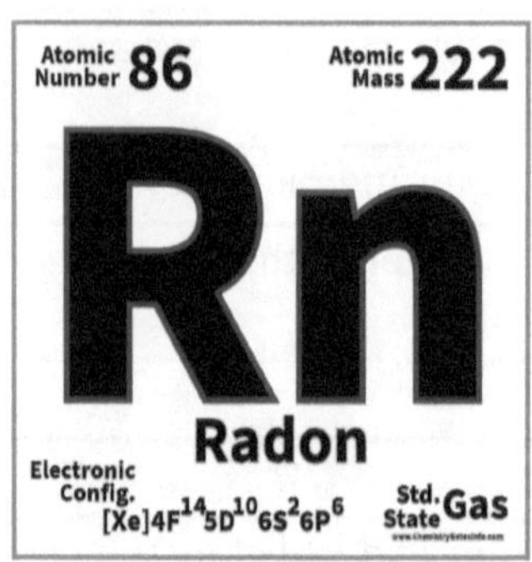

❖Radon is part of the noble gas group, meaning it is chemically inert and does not easily react with other elements.

❖Radon is a radioactive gas, and its most stable isotope, Radon-222, has a half-life of 3.8 days. It is one of the heaviest gases and is hazardous due to its radioactivity.

❖Radon is a colourless, odourless, and tasteless gas, which makes it difficult to detect without specialised equipment.

❖Radon is considered a major health risk because it can accumulate in buildings, especially basements, and inhaling radon over time increases the risk of lung cancer. It is the second leading cause of lung cancer after smoking.

❖ Radon is a decay product of uranium and thorium. It is continuously produced in the Earth's crust and released into the atmosphere through the breakdown of radioactive materials.

❖ Radon is the heaviest known gas at room temperature, with a density approximately 9.73 g/L, making it about nine times heavier than air.

❖ Radon was discovered in 1900 by Friedrich Ernst Dorn, a German physicist, while studying the decay of radium.

❖ Historically, radon was used in cancer therapy (radiotherapy) due to its radioactive properties, though its use in medicine has largely been discontinued due to safety concerns.

❖ Certain building materials, like granite, emit small amounts of radon. Homes built on radon-emitting rock formations are at higher risk of radon accumulation.

❖ Radon levels in homes are typically measured in picocuries per litre (pCi/L). EPA recommends mitigation for radon levels above 4pCi/L

❖ Proper ventilation in homes can reduce accumulation of radon gas.

PHYSICAL PROPERTIES:

Physical Property	Description
Atomic Symbol	Rn
Atomic Number	86
Atomic Mass	222 amu (for Radon-222, the most stable isotope)
State at Room Temp	Gas
Color	Colorless
Odor	Odourless
Density	9.73 g/L (at room temperature, making it the heaviest noble gas)
Melting Point	-71°C
Boiling Point	-61.7°C
Solubility in Water	Slightly soluble in water, about 230 cm³/L at 20°C
Conductivity	Does not conduct electricity or heat
Abundance in Earth	Trace amounts, mainly produced as a decay product of uranium in soil, rocks, and water

CHEMICAL PROPERTIES:

Chemical Property	Description
Reactivity	Radon is chemically inert under most conditions, like other noble gases
Oxidation States	Has a theoretical oxidation state of +2, though it rarely forms compounds
Electronegativity	Approximately 2.2 (Pauling scale), but it is not usually involved in chemical bonds
Reaction with Water	Slightly soluble in water but does not react chemically with it
Reaction with Acids	Radon does not react with acids
Reaction with Oxygen	Can form radon oxides in very rare, high-energy conditions
Flammability	Non-flammable
Toxicity	Extremely hazardous due to its radioactivity and inhalation risks
Biological Importance	No known biological role; prolonged exposure is dangerous to health, causing lung cancer risks
Isotopes	Radon has several isotopes, with Radon-222 being the most significant for environmental health

FRANCIUM

Hello rare element hunters! I'm Francium, the rarest of the alkali metals, and highly radioactive. You'll mostly find me in labs since I'm so unstable.

INTERESTING FACTS ABOUT FRANCIUM

❖Francium is one of the rarest naturally occurring elements, with an estimated less than 30 grams existing on Earth at any given time.

❖Francium is extremely radioactive, with a very short half-life. Its most stable isotope, Francium-223, has a half-life of only 22 minutes.

❖Francium was discovered by Marguerite Perey in 1939 and was named after France, her native country.

❖ As the heaviest alkali metal, francium has the largest atomic radius of all elements in Group 1 of the periodic table.

❖ Francium was the last naturally occurring alkali metal to be discovered, completing the alkali metal family.

❖ As an alkali metal, francium is highly reactive, especially with water, though this reactivity has rarely been observed directly due to its rarity and radioactivity.

❖ Francium is considered the most electropositive element, meaning it has a strong tendency to lose its single valence electron.

❖ Due to its short half-life, francium undergoes rapid radioactive decay, generating a significant amount of heat in the process. Francium salts have never been collected or studied in bulk because the element decays too quickly to form stable compounds.

❖ Francium is so rare that its total amount in the Earth's crust is less than a typical grain of sand in weight.

❖ Francium has no stable isotopes. The most important isotope, Francium-223, is produced as a decay product of actinium.

❖ Francium has no significant industrial applications due to its extreme rarity and radioactivity. It is primarily used for research purposes.

❖ Francium isotopes are so short-lived that studying the element in any meaningful amount is difficult and requires complex techniques.

PHYSICAL PROPERTIES:

Physical Property	Description
Atomic Symbol	Fr
Atomic Number	87
Atomic Mass	223 amu (for Francium-223, the most stable isotope)
State at Room Temp	Solid (metal, but difficult to observe due to its radioactivity)
Color	Likely silvery-gray, like other alkali metals, though its appearance is not well documented
Odor	Odorless
Density	Estimated to be around 2.48 g/cm^3
Melting Point	27°C (estimated, but rarely observed due to francium's short half-life)
Boiling Point	677°C (estimated)
Conductivity	Expected to be a good conductor of electricity, like other alkali metals
Abundance in Earth	Extremely rare, found in trace amounts in uranium and thorium ores

CHEMICAL PROPERTIES:

Chemical Property	Description
Reactivity	Highly reactive, especially with water and halogens, but this has rarely been observed
Oxidation States	+1 (like other alkali metals)
Electronegativity	0.7 (Pauling scale), making it the least electronegative element
Reaction with Water	Expected to react explosively with water to form francium hydroxide and hydrogen gas
Reaction with Acids	Reacts with acids to produce francium salts and hydrogen gas
Reaction with Oxygen	Likely forms francium oxide (Fr_2O) when exposed to air
Flammability	Not flammable, but highly reactive and hazardous due to radioactivity
Toxicity	Extremely hazardous due to its intense radioactivity
Biological Importance	No known biological role, and exposure would be dangerous due to radiation

RADIUM

Hi radiologists! I'm Radium, once used to glow in the dark and in cancer treatments. I'm radioactive and dangerous, so my use has shifted more to research.

INTERESTING FACTS ABOUT RADIUM

❖Radium was discovered in 1898 by the pioneering scientists Marie and Pierre Curie, who extracted it from uranium ore.

❖Radium is extremely radioactive, and it glows faintly in the dark due to the intense radiation it emits.

❖Radium was one of the first elements used in cancer treatment, specifically in radiotherapy, where its radiation was used to target and destroy cancer cells.

❖Radium was once widely used in luminous paints for watch dials, aircraft instruments, and clocks, but this practice was abandoned after it was discovered that prolonged exposure caused serious health issues.

❖ The "Radium Girls" were factory workers who applied radium-based paint to watches. Many of them suffered from radiation poisoning due to direct contact with radium.

❖ Radium undergoes radioactive decay to produce radon, a radioactive noble gas that poses a health risk when it accumulates in enclosed spaces.

❖ Radium is part of the alkaline earth metal group (Group 2) in the periodic table and is the heaviest member of this group.

❖ Like other alkaline earth metals, radium is highly reactive, especially with water, forming radium hydroxide and releasing hydrogen gas.

❖ When radium compounds are exposed to air, they glow with a bright, bluish light due to the continuous ionisation of the surrounding air caused by its radiation.

❖ Radium is typically found in trace amounts in uranium and thorium ores, particularly pitchblende.

❖ Radium exposure is extremely dangerous. Ingesting or inhaling radium can cause bone cancer due to its accumulation in bones.

PHYSICAL PROPERTIES:

Physical Property	Description
Atomic Symbol	Ra
Atomic Number	88
Atomic Mass	226 amu (for Radium-226, the most stable isotope)
State at Room Temp	Solid (metal)
Color	Silvery-white, but tarnishes quickly when exposed to air
Odor	Odourless
Density	5.5 g/cm^3
Melting Point	700°C
Boiling Point	1,737°C
Solubility in Water	Reacts with water, forming radium hydroxide (Ra(OH)$_2$) and releasing hydrogen gas
Conductivity	Good conductor of electricity and heat, similar to other metals
Abundance in Earth	Very rare, found in trace amounts in uranium ores

CHEMICAL PROPERTIES:

Chemical Property	Description
Reactivity	Highly reactive, especially with water and acids
Oxidation States	+2 (like other alkaline earth metals)
Electronegativity	0.9 (Pauling scale)
Reaction with Water	Reacts with water to form radium hydroxide and hydrogen gas
Reaction with Acids	Reacts with acids to produce radium salts and hydrogen gas
Reaction with Oxygen	Forms radium oxide (RaO) when exposed to air, causing it to tarnish quickly
Flammability	Non-flammable, but highly reactive
Toxicity	Extremely toxic due to its radioactivity; prolonged exposure can cause radiation sickness
Biological Importance	No known biological role; exposure can lead to serious health problems, including bone cancer
Isotopes	Radium has over 30 isotopes, with Radium-226 being the most common and longest-lived

ACTINIUM

Hey glow seekers! I'm Actinium, glowing with blue light due to my intense radioactivity. I'm used in radiation therapy and act as a gateway to the actinides.

INTERESTING FACTS ABOUT ACTINIUM

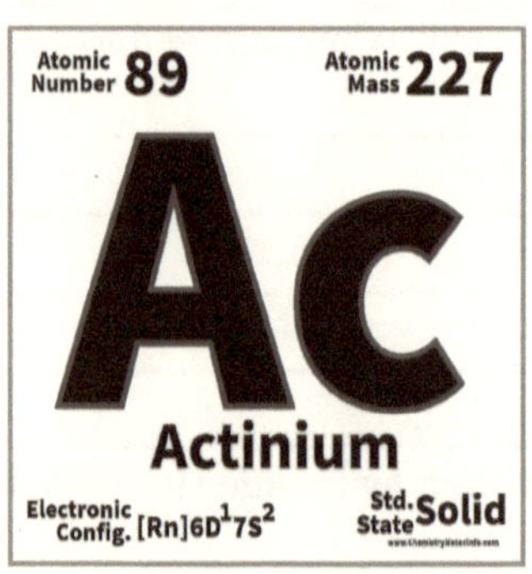

❖Actinium was the first element discovered that is solely radioactive, identified by Friedrich Oskar Giesel in 1899.

❖Actinium glows faintly blue in the dark due to the ionisation of air caused by its intense radioactivity.

❖The name "Actinium" comes from the Greek word aktinos, meaning "ray," in reference to the intense radiation it emits.

❖Actinium is extremely radioactive, with Actinium-227 being the most common isotope, having a half-life of 21.77 years.

❖ Actinium is found in trace amounts in uranium and thorium ores, usually alongside other radioactive elements.

❖ Actinium-225 is being researched and used in targeted alpha-particle cancer therapies, showing promise in treating some forms of cancer.

❖ Actinium is the first element in the actinide series, a group of 15 elements that are mostly radioactive and have similar properties.

❖ Due to its rarity and intense radioactivity, actinium has limited industrial applications, mostly confined to research and medical fields.

❖ Chemically, actinium is quite similar to lanthanum, which is why they are often difficult to separate from each other.

❖ Actinium decays into radium and then into radon, contributing to the natural occurrence of this radioactive gas.

❖ Actinium can be produced synthetically by bombarding radium with neutrons, although this is done mainly for research.

❖ Actinium is so radioactive that even small samples generate significant heat, enough to make the material warm to the touch.

❖ Actinium is an efficient source of alpha particles, making it useful for radiation therapy and in nuclear research.

PHYSICAL PROPERTIES:

Physical Property	Description
Atomic Symbol	Ac
Atomic Number	89
Atomic Mass	227 amu (for Actinium-227, the most stable isotope)
State at Room Temp	Solid (metal)
Color	Silvery-white
Odor	Odourless
Density	10.07 g/cm³
Melting Point	1,050°C
Boiling Point	3,198°C
Solubility in Water	Slightly soluble, but mostly reacts with water to form actinium hydroxide
Conductivity	Good conductor of heat and electricity
Abundance in Earth	Found in trace amounts in uranium and thorium ores

CHEMICAL PROPERTIES:

Chemical Property	Description
Reactivity	Highly reactive, especially with oxygen, halogens, and acids
Oxidation States	+3 (dominant oxidation state in compounds)
Electronegativity	1.1 (Pauling scale)
Reaction with Water	Reacts slowly with water, forming actinium hydroxide
Reaction with Acids	Reacts readily with most acids to form soluble actinium salts
Reaction with Oxygen	Reacts with oxygen to form a protective oxide layer (Ac_2O_3)
Flammability	Non-flammable, but highly reactive
Toxicity	Extremely toxic due to its radioactivity; exposure can cause radiation sickness and cancer
Biological Importance	No known biological role; exposure is hazardous
Isotopes	Actinium has several isotopes, with Actinium-227 being the most stable and important

THORIUM

Hello nuclear pioneers! I'm Thorium, a potential future fuel for nuclear reactors. I'm abundant and efficient, offering a cleaner path to nuclear energy.

INTERESTING FACTS ABOUT THORIUM

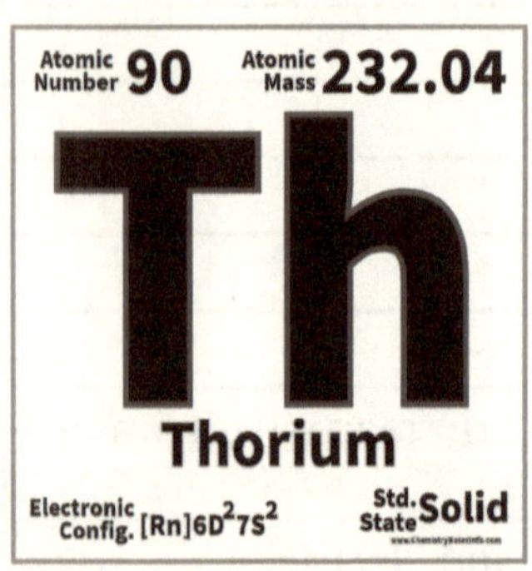

❖Thorium is named after Thor, the Norse god of thunder, due to its powerful and energetic nature.

❖Thorium is being explored as a potential alternative to uranium in nuclear reactors, as it is more abundant and produces less long-lived radioactive waste.

❖Thorium was discovered by Swedish chemist Jöns Jakob Berzelius in 1828 in a mineral called thorite.

❖Thorium emits a faint blue glow when it decays, caused by the ionisation of air around it.

❖ Thorium is about three to four times more abundant in the Earth's crust than uranium, making it an attractive option for future energy production.

❖ Thorium was once commonly used in gas lamp mantles to produce a bright, white light when heated. This practice has been largely discontinued due to the radioactivity of thorium.

❖ Unlike some other radioactive elements, thorium is considered relatively low in toxicity when it is in its metallic form, though it is still hazardous in its compounds and dust form.

❖ The most stable isotope, Thorium-232, has an incredibly long half-life of 14 billion years, making it nearly as old as the universe itself.

❖ Due to its high melting point, thorium is used in alloys for high-temperature applications such as in aircraft engines and certain types of welding.

❖ In nuclear reactors, thorium can be used to breed Uranium-233, a fissile material that can be used as nuclear fuel.

❖ Although thorium has potential in nuclear energy, current reactor designs and infrastructure are not yet optimised for thorium-based fuels, limiting its current use.

PHYSICAL PROPERTIES:

Property	Description
Atomic Symbol	Th
Atomic Number	90
Atomic Mass	232.04 amu
State at Room Temp	Solid (metal)
Color	Silvery white, but tarnishes to gray or black when exposed to air
Odor	Odorless
Density	11.7 g/cm^3
Melting Point	1,750°C
Boiling Point	4,790°C
Solubility in Water	Insoluble in water, but can react with water at elevated temperatures
Conductivity	Good conductor of heat and electricity
Abundance in Earth	Relatively abundant, about 3 to 4 times more common than uranium

CHEMICAL PROPERTIES:

Chemical Property	Description
Reactivity	Moderately reactive, especially at high temperatures
Oxidation States	+4 (dominant oxidation state in compounds)
Electronegativity	1.3 (Pauling scale)
Reaction with Water	Reacts very slowly with water, but faster at high temperatures
Reaction with Acids	Dissolves in strong acids, particularly nitric acid
Reaction with Oxygen	Forms a protective oxide layer (ThO_2) when exposed to air, preventing further oxidation
Flammability	Non-flammable, but thorium powder can ignite in air
Toxicity	Thorium compounds and dust can be hazardous if inhaled or ingested due to radioactivity
Biological Importance	No known biological role; exposure to thorium dust can cause radiation poisoning
Isotopes	Thorium-232 is the most common and stable isotope; it is used as a potential nuclear fuel

PROTACTINIUM

Hi researchers! I'm Protactinium, a rare and highly radioactive element. I'm mostly found in research labs, helping scientists understand nuclear reactions.

INTERESTING FACTS ABOUT PROTACTINIUM

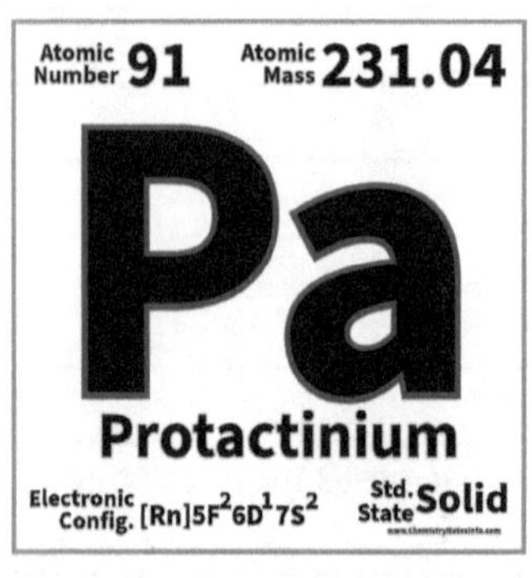

❖The name "Protactinium" comes from the Greek words "protos" (meaning first) and "actinium," because it is the precursor to actinium in radioactive decay chains.

❖Protactinium was discovered in 1913 by Kasimir Fajans and Otto Gohring, but its properties were more fully explored in 1917 by Lise Meitner and Otto Hahn.

❖Protactinium is one of the rarest naturally occurring elements, found in trace amounts in uranium ores.

❖ Protactinium is a highly radioactive element, with the isotope Protactinium-231 being most stable with a half-life of 32,760 years.

❖ Protactinium is part of the uranium-238 decay chain, decaying into actinium and eventually into stable lead.

❖ Due to its scarcity and radioactivity, protactinium has no significant commercial applications, but it is of interest in scientific research, especially in nuclear studies.

❖ It is difficult to isolate in its pure form due to its reactivity and the fact that it is usually found in very low concentrations in nature.

❖ Protactinium is a dense, silvery-grey metal with a high melting point, similar in appearance to uranium.

❖ Because of its radioactivity, protactinium is hazardous and must be handled with special precautions to avoid radiation exposure.

❖ During the 1940s and 1950s, protactinium was studied as part of early nuclear research, though its use was largely abandoned in favour of more abundant elements like uranium.

❖ While not used in current reactors, protactinium's nuclear properties suggest it could be used in certain types of nuclear reactions, especially for production of U-233 in thorium reactors.

PHYSICAL PROPERTIES:

Physical Property	Description
Atomic Symbol	Pa
Atomic Number	91
Atomic Mass	231.04 amu
State at Room Temp	Solid (metal)
Color	Silvery metallic, but tarnishes in air
Odor	Odourless
Density	15.37 g/cm³
Melting Point	1,572°C
Boiling Point	4,000°C (estimated)
Solubility in Water	Insoluble in water
Conductivity	Good conductor of heat and electricity
Abundance in Earth	Extremely rare, found in trace amounts in uranium ores

CHEMICAL PROPERTIES:

Chemical Property	Description
Reactivity	Reacts with oxygen and acids, forming oxides and other compounds
Oxidation States	+5 (most stable), +4, +3
Electronegativity	1.5 (Pauling scale)
Reaction with Water	Reacts slowly with water
Reaction with Acids	Dissolves in most acids, forming protactinium salts
Reaction with Oxygen	Forms a protective oxide layer (Pa_2O_5) when exposed to air
Flammability	Non-flammable, but its powder form can be hazardous due to radioactivity
Toxicity	Extremely toxic due to radioactivity; exposure can lead to radiation sickness and cancer
Biological Importance	No known biological role; exposure can be hazardous
Isotopes	Protactinium-231 is the most common and stable isotope, used in nuclear research

URANIUM

Hello power generators! I'm Uranium, the fuel behind nuclear power and atomic bombs. I'm a heavy hitter when it comes to energy—both peaceful and powerful.

INTERESTING FACTS ABOUT URANIUM

❖Uranium was named after the planet Uranus, which had been discovered just eight years before the element.

❖Uranium was first discovered by the German chemist Martin Heinrich Klaproth in 1789 in the mineral pitchblende.

❖The discovery of nuclear fission in uranium by Otto Hahn and Fritz Strassmann in 1938 paved the way for the development of nuclear energy and weapons.

❖ Uranium is the primary fuel used in nuclear power plants to generate electricity through fission.

❖ Uranium is a very dense metal, about 1.7 times denser than lead, making it useful in applications requiring high density, such as armour-piercing ammunition.

❖ Uranium was once used to make a type of glassware that glows green under ultraviolet light, known as uranium glass or Vaseline glass.

❖ Uranium naturally occurs as three isotopes: U-238 (the most abundant), U-235 (used in nuclear reactors), and U-234.

❖ Uranium-235 is the only naturally occurring isotope that is fissionable, making it crucial for nuclear reactors and atomic bombs.

❖ Although relatively rare compared to elements like iron or aluminium, uranium is more common than gold or silver in the Earth's crust.

❖ Uranium undergoes radioactive decay to form radium and eventually lead, a process that takes billions of years.

❖ The process of "enriching" uranium involves increasing the proportion of Uranium-235 for use in nuclear reactors or weapons.

❖ Uranium-235 was used in the "Little Boy" atomic bomb dropped on Hiroshima, marking its historical significance in warfare.

❖ Uranium is both chemically toxic and radioactive, posing significant health risks if ingested, inhaled, or handled improperly.

PHYSICAL PROPERTIES:

Physical Property	Description
Atomic Symbol	U
Atomic Number	92
Atomic Mass	238.03 amu
State at Room Temp	Solid (metal)
Color	Silvery-white, but tarnishes to black when exposed to air
Odor	Odourless
Density	18.95 g/cm^3
Melting Point	1,132°C
Boiling Point	4,131°C
Solubility in Water	Insoluble in water
Conductivity	Good conductor of heat and electricity
Abundance in Earth	Fairly common, found in various ores such as pitchblende and uraninite

CHEMICAL PROPERTIES:

Chemical Property	Description
Reactivity	Reacts with oxygen, halogens, and acids
Oxidation States	+6 (most stable), +5, +4, +3
Electronegativity	1.38 (Pauling scale)
Reaction with Water	Reacts slowly with water at room temperature, but more quickly in hot water
Reaction with Acids	Dissolves in acids such as nitric acid, forming uranyl ions (UO_2^{2+})
Reaction with Oxygen	Forms uranium oxide (UO_2) when exposed to air, which further oxidizes to U_3O_8
Flammability	Non-flammable, but finely divided uranium powder can ignite spontaneously in air
Toxicity	Highly toxic due to both chemical toxicity and radioactivity; can cause kidney damage and cancer
Biological Importance	No known biological role; exposure is hazardous
Isotopes	Uranium-238 (most abundant), Uranium-235 (fissile), Uranium-234

NEPTUNIUM

Hi experimenters! I'm Neptunium, a byproduct of uranium reactions. I'm used in research and have potential applications in space exploration due to my radioactive energy.

INTERESTING FACTS ABOUT NEPTUNIUM

❖Neptunium is named after the planet Neptune, following uranium, which was named after Uranus, the next planet in the solar system.

❖Neptunium was the first element to be discovered that is heavier than uranium, making it the first "transuranic" element.

❖Neptunium was discovered by Edwin McMillan and Philip H. Abelson in 1940 at the University of California, Berkeley, by bombarding uranium with neutrons.

❖ Neptunium is highly radioactive, with no stable isotopes, though Neptunium-237 is the longest-lived isotope, with a half-life of about 2.14 million years.

❖ Neptunium is a byproduct of nuclear reactors and is formed during the fission of uranium and plutonium.

❖ While not used in current nuclear weapons, Neptunium-237 could theoretically be used to create a nuclear bomb due to its ability to undergo fission.

❖ Neptunium is chemically and radiologically toxic, with significant risks to human health if inhaled or ingested.

❖ Neptunium isotopes, like Plutonium-238, are sometimes used in radioisotope thermoelectric generators (RTGs) to power space probes and satellites.

❖ Although primarily man-made, tiny amounts of neptunium can form naturally in uranium ores due to radioactive decay.

❖ Neptunium belongs to the actinide series of the periodic table, a group of 15 elements known for their similar properties and radioactive nature.

❖ Neptunium-237 decays into americium, another actinide, through alpha decay over millions of years.

PHYSICAL PROPERTIES:

Physical Property	Description
Atomic Symbol	Np
Atomic Number	93
Atomic Mass	237.05 amu
State at Room Temp	Solid (metal)
Color	Silvery metallic
Odor	Odourless
Density	20.45 g/cm³
Melting Point	640°C
Boiling Point	4,174°C
Solubility in Water	Insoluble in water
Conductivity	Good conductor of heat and electricity
Abundance in Earth	Extremely rare, found in trace amounts in uranium ores

CHEMICAL PROPERTIES:

Chemical Property	Description
Reactivity	Highly reactive, especially with oxygen and acids
Oxidation States	+3, +4, +5, +6 (with +5 being the most stable)
Electronegativity	1.36 (Pauling scale)
Reaction with Water	Slowly reacts with water
Reaction with Acids	Reacts with acids, dissolving to form neptunium salts
Reaction with Oxygen	Reacts with oxygen to form neptunium oxides
Flammability	Non-flammable, but neptunium powder can ignite spontaneously in air
Toxicity	Highly toxic both chemically and radiologically; can cause severe radiation poisoning
Biological Importance	No known biological role; exposure is dangerous
Isotopes	Neptunium-237 (most stable), Neptunium-239 (precursor to plutonium)

PLUTONIUM

Hello nuclear experts! I'm Plutonium, a key fuel for nuclear reactors and weapons. Powerful and dangerous, I've been central to both energy generation and military history.

INTERESTING FACTS ABOUT PLUTONIUM

❖Plutonium is named after the dwarf planet Pluto, continuing the tradition of naming transuranic elements after planets.

❖Plutonium was first synthesised by Glenn T. Seaborg, Arthur Wahl, & Joseph Kennedy at the University of California, Berkeley in 1940.

❖Plutonium-239, a fissile isotope, is used as a primary material in nuclear weapons, including bomb dropped on Nagasaki in 1945.

❖Plutonium-239 is also used as fuel in some nuclear reactors, particularly in fast breeder reactors, which generate more plutonium than they consume.

❖ Plutonium is highly radioactive, emitting alpha particles that pose a significant health hazard if ingested or inhaled.

❖ The most common isotopes are Plutonium-239 and Plutonium-240. Plutonium-238 is used in radioisotope thermoelectric generators (RTGs) for space missions.

❖ Due to its radioactivity, plutonium generates heat as it decays, making it useful in RTGs that power space probes like Voyager and the Curiosity rover on Mars.

❖ Freshly prepared plutonium has a silvery appearance, but it quickly tarnishes to a dark colour when exposed to air.

❖ Plutonium does not occur naturally in significant quantities. It is produced in nuclear reactors by bombarding U-238 with neutrons.

❖ Plutonium exists in six different crystalline forms (allotropes), each with different densities and properties, making it highly complex to work with.

❖ Plutonium's discovery was initially kept secret due to its role in the Manhattan Project and was only made public after World War II.

❖ It is a significant component of nuclear waste from reactors, with long-lived isotopes that remain radioactive for thousands of years.

PHYSICAL PROPERTIES:

Physical Property	Description
Atomic Symbol	Pu
Atomic Number	94
Atomic Mass	244 amu
State at Room Temp	Solid (metal)
Color	Silvery white, tarnishes to yellow or olive green
Odor	Odourless
Density	19.86 g/cm^3
Melting Point	640°C
Boiling Point	3,228°C
Solubility in Water	Insoluble in water
Conductivity	Poor conductor of heat and electricity compared to other metals
Abundance in Earth	Extremely rare, almost entirely man-made

CHEMICAL PROPERTIES:

Chemical Property	Description
Reactivity	Highly reactive, especially with oxygen, halogens, and acids
Oxidation States	+3, +4, +5, +6 (with +4 being the most stable)
Electronegativity	1.28 (Pauling scale)
Reaction with Water	Reacts slowly with water
Reaction with Acids	Reacts with most acids to form plutonium salts
Reaction with Oxygen	Reacts with oxygen to form oxides (e.g., PuO_2, Pu_2O_3)
Flammability	Non-flammable, but finely divided plutonium powder can ignite spontaneously in air
Toxicity	Extremely toxic, both chemically and radiologically; exposure can cause severe health problems
Biological Importance	No known biological role; exposure is dangerous
Isotopes	Plutonium-239 (fissile), Plutonium-238 (used in space exploration), Plutonium-240

AMERICIUM

Hi smoke detector owners! I'm Americium, found in most household smoke detectors. My radioactive properties help keep your home safe from fires.

INTERESTING FACTS ABOUT AMERICIUM

❖Americium is named after the Americas, following the naming convention of elements after continents (similar to europium).

❖Americium was discovered by Glenn T. Seaborg and his team in 1944 as part of the Manhattan Project while working on plutonium chemistry.

❖Americium-241 is widely used in household smoke detectors due to its ability to ionise air and detect smoke particles.

❖Americium is part of the actinide series and was the first element in the group after curium to be synthesised.

❖ Americium is highly radioactive, with Americium-241 having a half-life of about 432 years. It primarily decays by emitting alpha particles.

❖ Like most transuranic elements, americium does not occur naturally. It is produced by bombarding plutonium with neutrons in a nuclear reactor.

❖ Due to its intense radioactivity, small amounts of americium will glow faintly in the dark.

❖ The radioactivity of americium poses serious health risks if inhaled or ingested, leading to possible radiation poisoning and cancer.

❖ Americium-241 is often combined with beryllium to form neutron sources used in industrial gauging and medical applications.

❖ Americium can undergo fission, but it is less efficient compared to plutonium or uranium, so it is rarely used as a nuclear fuel.

❖ The discovery of americium was first publicly announced on a children's radio show in 1945, making it one of the most unusual scientific announcements in history.

❖ Americium emits alpha particles, which are relatively weak and stopped by paper or skin, but can cause severe damage if they enter the body.

❖ Americium is a significant component of nuclear waste.

PHYSICAL PROPERTIES:

Physical Property	Description
Atomic Symbol	Am
Atomic Number	95
Atomic Mass	243 amu
State at Room Temp	Solid (metal)
Color	Silvery-white metallic
Odor	Odourless
Density	12 g/cm^3
Melting Point	1,176°C
Boiling Point	2,011°C
Solubility in Water	Insoluble in water
Conductivity	Poor conductor of electricity compared to other metals
Abundance in Earth	Extremely rare; almost entirely man-made

CHEMICAL PROPERTIES:

Chemical Property	Description
Reactivity	Reacts with oxygen, acids, and halogens
Oxidation States	+2, +3, +4, +5, +6 (with +3 being the most stable)
Electronegativity	1.3 (Pauling scale)
Reaction with Water	Reacts slowly with water
Reaction with Acids	Dissolves in acids, forming americium salts
Reaction with Oxygen	Forms oxides (Am_2O_3, AmO_2) when exposed to air
Flammability	Non-flammable, but finely divided americium powder can ignite spontaneously
Toxicity	Extremely toxic both chemically and radiologically; causes radiation sickness and cancer
Biological Importance	No known biological role; exposure is harmful
Isotopes	Americium-241 (used in smoke detectors), Americium-243 (most stable)

CURIUM

Hey explorers! I'm Curium, used in space exploration as a power source for spacecraft. My radioactivity helps power missions far beyond Earth.

INTERESTING FACTS ABOUT CURIUM

❖Curium was named in honour of Marie and Pierre Curie, pioneers in radioactivity research.

❖Curium was first synthesised in 1944 by scientists Glenn T. Seaborg, Albert Ghiorso, and James, working as part of the Manhattan Project.

❖Curium is a highly radioactive element, with most of its isotopes emitting alpha particles. It poses significant health risks if inhaled or ingested.

❖ Unlike other actinides, curium glows in the dark due to its radioactivity, emitting a faint purple or blue glow from excited electrons.

❖ Curium isotopes, particularly Curium-244, have been used in radioisotope thermoelectric generators (RTGs) to power satellites and space probes.

❖ Curium is not found naturally in significant amounts but is produced in nuclear reactors as a byproduct of plutonium and uranium fission.

❖ Due to its strong radioactivity, curium generates significant heat as it decays, which is useful in applications like RTGs.

❖ Curium's radioactivity makes it highly toxic. Exposure to curium can cause radiation burns, lung cancer, and other health issues.

❖ Curium-247 is one of the most stable isotopes of curium, with a half-life of about 15.6 million years, making it a long-term environmental hazard.

❖ Curium is highly reactive and readily combines with oxygen, water, and other elements, forming compounds like curium oxide (Cm_2O_3).

PHYSICAL PROPERTIES:

Physical Property	Description
Atomic Symbol	Cm
Atomic Number	96
Atomic Mass	247 amu
State at Room Temp	Solid (metal)
Color	Silvery-white
Odor	Odourless
Density	13.52 g/cm^3
Melting Point	1,345°C
Boiling Point	3,110°C
Solubility in Water	Insoluble in water
Conductivity	Good conductor of electricity and heat
Abundance in Earth	Extremely rare, almost entirely man-made

CHEMICAL PROPERTIES:

Chemical Property	Description
Reactivity	Reacts with oxygen, acids, and halogens
Oxidation States	+3, +4 (with +3 being the most stable)
Electronegativity	1.3 (Pauling scale)
Reaction with Water	Reacts slowly with water
Reaction with Acids	Dissolves in acids to form curium salts
Reaction with Oxygen	Reacts with oxygen to form oxides (Cm_2O_3, CmO_2)
Flammability	Non-flammable, but curium powder can ignite spontaneously in air
Toxicity	Extremely toxic both chemically and radiologically; causes severe health risks
Biological Importance	No known biological role; harmful to living organisms
Isotopes	Curium-242 (used in space exploration), Curium-244 (used in RTGs), Curium-247 (long-lived)

BERKELIUM

Hello scientists! I'm Berkelium, created in labs and used in research. I'm helping scientists push the boundaries of chemistry and nuclear physics.

INTERESTING FACTS ABOUT BERKELIUM

❖Berkelium was named after the city of Berkeley, where it was first synthesised at the University of California, Berkeley.

❖Berkelium was discovered in 1949 by scientists Glenn T. Seaborg, Albert Ghiorso, and Stanley G. Thompson by bombarding americium-241 with alpha particles.

❖Berkelium does not occur naturally and is entirely synthesised in nuclear reactors or particle accelerators.

❖Berkelium is highly radioactive, and its most common isotope, Berkelium-249, emits beta particles and has a half-life of about 330 days.

❖ Berkelium is produced in extremely small amounts, making it one of the rarest elements. It is expensive to produce and difficult to isolate.

❖ Berkelium is primarily used for research purposes, particularly in the synthesis of heavier elements like tennessine (element 117).

❖ It is a soft, silvery metal, but it tarnishes quickly when exposed to air.

❖ Due to its radioactivity, berkelium generates heat as it decays, although not enough for practical energy applications like plutonium or curium.

❖ Berkelium has no significant commercial or industrial uses due to its scarcity and radioactivity, but it is important in nuclear science.

❖ Like other actinides, berkelium shares chemical similarities with elements like curium and californium, especially in its +3 oxidation state.

❖ Due to its radioactivity and the difficulty of production, handling and isolating berkelium requires highly specialised equipment and safety protocols.

❖ In 2009, a rare 22 milligrams of berkelium was produced at Oak Ridge National Laboratory (USA), and it was later used in Russia to help discover the superheavy element tennessine (Ts, element 117). This made berkelium essential in extending the periodic table.

PHYSICAL PROPERTIES:

Physical Property	Description
Atomic Symbol	Bk
Atomic Number	97
Atomic Mass	247 amu
State at Room Temp	Solid (metal)
Color	Silvery white
Odor	Odourless
Density	14.78 g/cm^3
Melting Point	986°C
Boiling Point	2,623°C
Solubility in Water	Insoluble in water
Conductivity	Moderate conductor of electricity and heat
Abundance in Earth	Extremely rare, entirely synthesised

CHEMICAL PROPERTIES:

Chemical Property	Description
Reactivity	Reacts with oxygen, acids, and halogens
Oxidation States	+3 (most stable), +4
Electronegativity	1.3 (Pauling scale)
Reaction with Water	Reacts slowly with water
Reaction with Acids	Reacts with acids to form berkelium salts
Reaction with Oxygen	Forms oxides like Bk_2O_3 when exposed to air
Flammability	Non-flammable, but finely divided berkelium powder can ignite
Toxicity	Highly toxic and radioactive, causing severe health risks
Biological Importance	No known biological role; harmful to organisms due to its radioactivity
Isotopes	Berkelium-247 (stable for research), Berkelium-249 (most common and radioactive)

CALIFORNIUM

Greetings experimenters! I'm Californium, a super radioactive element used in neutron radiation. I'm vital for finding gold and silver in mining, and for starting nuclear reactors.

INTERESTING FACTS ABOUT CALIFORNIUM

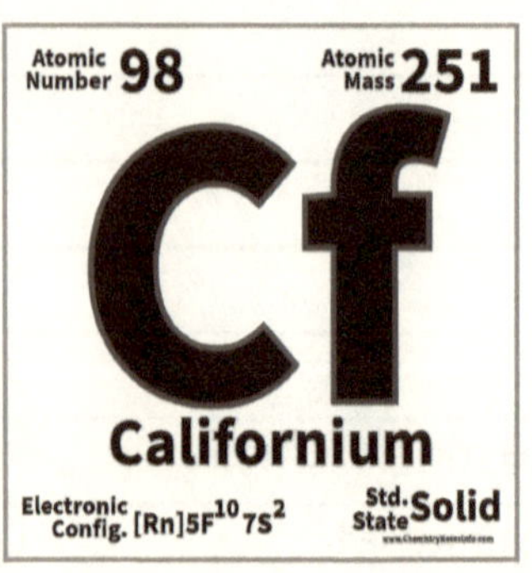

❖Californium is named after the state of California, where it was first synthesised at the University of California, Berkeley, in 1950.

❖Californium was discovered by scientists Stanley G. Thompson, Kenneth Street Jr., Albert Ghiorso, and Glenn T. Seaborg by bombarding curium-242 with alpha particles.

❖Californium-252 is known for its ability to emit neutrons, making it useful in neutron radiography, neutron activation analysis, and starting nuclear reactors.

❖ Californium is extremely radioactive, especially its isotope californium-252, which has a half-life of about 2.645 years.

❖ Californium-252 has medical applications, particularly in treating certain types of cancer through neutron therapy, as it can target tumours with its neutron emission.

❖ Californium is used in oil exploration to help determine the composition of geological formations through neutron analysis.

❖ C-252 is used to detect hidden explosives, landmines, & nuclear waste due to its ability to penetrate materials with neutrons.

❖ Californium's strong neutron emission has made it useful in nuclear research, especially in the study of other transuranic elements and for creating superheavy elements.

❖ Californium is very expensive and difficult to produce. The cost of californium-252 can exceed $27 million per gram.

❖ Californium has been used in the synthesis of heavier elements, including elements like tennessine (element 117), through particle bombardment experiments.

❖ Californium, like many transuranic elements, generates heat through radioactive decay.

PHYSICAL PROPERTIES:

Physical Property	Description
Atomic Symbol	Cf
Atomic Number	98
Atomic Mass	251 amu
State at Room Temp	Solid (metal)
Color	Silvery-white
Odor	Odourless
Density	15.1 g/cm³
Melting Point	900°C
Boiling Point	Unknown (estimated > 1,740°C)
Solubility in Water	Insoluble in water
Conductivity	Moderate conductor of electricity and heat
Abundance in Earth	Extremely rare, entirely synthetic

CHEMICAL PROPERTIES:

Chemical Property	Description
Reactivity	Reacts with oxygen, acids, and halogens
Oxidation States	+3 (most stable), +4
Electronegativity	1.3 (Pauling scale)
Reaction with Water	Reacts slowly with water
Reaction with Acids	Dissolves in acids to form californium salts
Reaction with Oxygen	Forms oxides like Cf_2O_3
Flammability	Non-flammable, but powdered californium can ignite
Toxicity	Extremely toxic and radioactive; can cause severe health risks
Biological Importance	No known biological role; harmful due to its radioactivity
Isotopes	Californium-249 (used for research), Californium-252 (used in neutron emission applications)

EINSTEINIUM

Hi pioneers! I'm Einsteinium, named after the genius himself. I'm synthetic and highly radioactive, found mostly in research for nuclear science.

INTERESTING FACTS ABOUT EINSTEINIUM

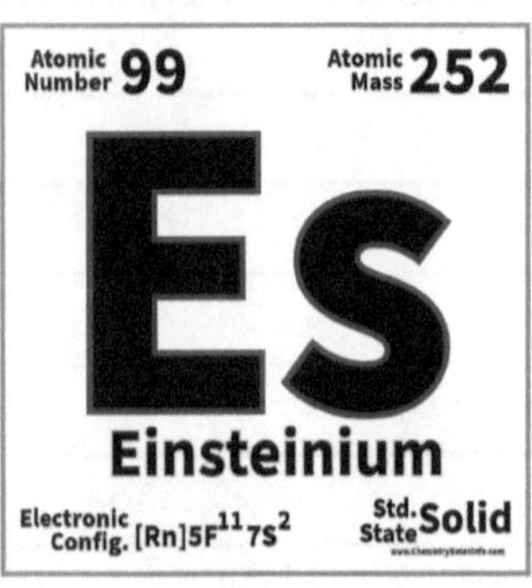

❖Einsteinium was named in honour of the famous physicist Albert Einstein, recognising his contributions to science.

❖Einsteinium was first identified in the debris of the first hydrogen bomb test, known as "Ivy Mike," in 1952 by a team of scientists from the University of California, Berkeley.

❖Einsteinium is a synthetic element, not found naturally on Earth, and is produced in small quantities in nuclear reactors or during nuclear explosions.

❖ Einsteinium is intensely radioactive, with its most common isotope, E-253, having a half-life of 20.5 days and emitting alpha particles.

❖ Einsteinium is one of the rarest elements on Earth, with only micrograms produced annually, making it extremely costly to synthesise.

❖ Due to its rarity and radioactivity, einsteinium has very limited practical uses, primarily being used for basic scientific research on heavy elements.

❖ Einsteinium decays into berkelium and other elements as it emits radiation, making it challenging to study for extended periods.

❖ As an actinide, einsteinium shares similarities with other heavy elements in this series, particularly in its +3 oxidation state.

❖ Einsteinium glows faintly in the dark due to the energy released from its radioactive decay.

❖ Einsteinium generates significant heat due to its radioactive decay, though this heat is not used for any practical energy applications.

❖ Einsteinium has been used in the synthesis of heavier elements, such as mendelevium (element 101), through particle bombardment.

❖ Einsteinium is difficult to work with due to its radioactivity.

PHYSICAL PROPERTIES:

Physical Property	Description
Atomic Symbol	Es
Atomic Number	99
Atomic Mass	252 amu
State at Room Temp	Solid (metal)
Color	Silvery white
Odor	Odourless
Density	8.84 g/cm^3
Melting Point	860°C
Boiling Point	Unknown
Solubility in Water	Insoluble in water
Conductivity	Moderate conductor of electricity and heat
Abundance in Earth	Extremely rare, entirely synthetic

CHEMICAL PROPERTIES:

Chemical Property	Description
Reactivity	Reacts with oxygen, acids, and halogens
Oxidation States	+3 (most stable), +2
Electronegativity	1.3 (Pauling scale)
Reaction with Water	Reacts slowly with water
Reaction with Acids	Dissolves in acids to form einsteinium salts
Reaction with Oxygen	Forms oxides like Es_2O_3
Flammability	Non-flammable, but einsteinium powder can ignite under certain conditions
Toxicity	Extremely toxic and radioactive, causing severe health risks
Biological Importance	No known biological role; highly harmful to living organisms due to its radioactivity
Isotopes	Einsteinium-253 (most common, highly radioactive)

FERMIUM

Hello physicists! I'm Fermium, created in nuclear explosions and used in research. I help scientists explore the far reaches of the periodic table.

INTERESTING FACTS ABOUT FERMIUM

❖Fermium was named in honour of the Italian physicist Enrico Fermi, who is best known for his work on nuclear reactors and for his contributions to nuclear physics.

❖Fermium was first identified in the debris of the "Ivy Mike" hydrogen bomb test in 1952, along with other heavy elements like einsteinium.

❖Fermium is a synthetic element and does not occur naturally on Earth. It is produced in nuclear reactors or during high-energy nuclear explosions.

❖ Fermium is highly radioactive, with its most common isotope, Fermium-257, having a half-life of 100.5 days and decaying by alpha emission.

❖ Only microgram quantities of fermium have ever been synthesised, making it extremely rare and difficult to study.

❖ Due to its scarcity & radioactivity, fermium has no known practical or commercial applications and is used only for scientific research.

❖ Fermium has been used in the study of nuclear fission and the behaviour of transuranium elements in extreme conditions.

❖ Fermium decays into lighter elements, primarily californium, through alpha emission, making it an interesting element to study for nuclear decay chains.

❖ Like other elements in the actinide series, fermium commonly exhibits a +3 oxidation state, although other oxidation states may also exist.

❖ Relatively short half-life of fermium isotopes limits time scientists have to conduct experiments, making its study more complex.

❖ Enrico Fermi, after whom fermium is named, was one of the pioneers of the Manhattan Project and played a crucial role in the development of nuclear energy.

PHYSICAL PROPERTIES:

Physical Property	Description
Atomic Symbol	Fm
Atomic Number	100
Atomic Mass	257 amu
State at Room Temp	Solid (metal)
Color	Unknown (likely silvery like other actinides)
Odor	Odourless
Density	Unknown (due to scarcity, physical properties are difficult to measure)
Melting Point	Estimated around 1,527°C
Boiling Point	Unknown
Solubility in Water	Insoluble in water
Conductivity	Likely a moderate conductor of electricity and heat, similar to other actinides
Abundance in Earth	Extremely rare, entirely synthetic

CHEMICAL PROPERTIES:

Chemical Property	Description
Reactivity	Reacts with oxygen, acids, and halogens
Oxidation States	+3 (most stable), +2
Electronegativity	1.3 (Pauling scale)
Reaction with Water	Reacts slowly with water
Reaction with Acids	Dissolves in acids to form fermium salts
Reaction with Oxygen	Forms oxides, likely similar to Fm_2O_3
Flammability	Non-flammable, but may ignite under specific conditions
Toxicity	Extremely toxic and radioactive, causing severe health risks
Biological Importance	No known biological role; hazardous to living organisms due to its intense radioactivity
Isotopes	Fermium-257 (most common), Fermium-255, Fermium-258 (used in research)

MENDELEVIUM

Hi element hunters! I'm Mendelevium, named after the creator of the periodic table. I'm man-made and used mainly in scientific research on heavy elements.

INTERESTING FACTS ABOUT MENDELEVIUM

❖Mendelevium is named in honour of Dmitri Mendeleev, the Russian chemist who developed the periodic table of elements.

❖Mendelevium was discovered in 1955 by a team of scientists led by Albert Ghiorso, Stanley Thompson, and Glenn T. Seaborg by bombarding einsteinium-253 with alpha particles at the University of California, Berkeley.

❖Mendelevium was the first element to be synthesised and identified on an atom-by-atom basis due to its highly limited production.

❖ Mendelevium does not occur naturally and is produced synthetically in particle accelerators.

❖ Mendelevium is radioactive, with its most stable isotope, Mendelevium-258, having a half-life of 51.5 days.

❖ Due to its radioactivity and the difficulty of producing it in significant quantities, mendelevium has no practical applications beyond basic scientific research.

❖ Mendelevium belongs to the actinide series of elements, which consists of heavy, mostly radioactive elements.

❖ Only microgram quantities of mendelevium have been produced, making it one of the rarest elements on Earth.

❖ Mendelevium's chemical properties are similar to other actinides, particularly those of americium and curium.

❖ Mendelevium is synthesised by bombarding einsteinium with alpha particles in a particle accelerator, producing atoms of mendelevium through nuclear fusion.

❖ The naming of mendelevium honours Dmitri Mendeleev's pivotal role in organising the known elements into a systematic table, which is still in use today.

PHYSICAL PROPERTIES:

Physical Property	Description
Atomic Symbol	Md
Atomic Number	101
Atomic Mass	258 amu
State at Room Temp	Solid (metal)
Color	Unknown (likely silvery or metallic like other actinides)
Odor	Odourless
Density	Unknown (due to scarcity, physical properties are difficult to measure)
Melting Point	Estimated around 827°C
Boiling Point	Unknown
Solubility in Water	Insoluble in water
Conductivity	Likely a moderate conductor of electricity and heat, similar to other actinides
Abundance in Earth	Extremely rare, entirely synthetic

CHEMICAL PROPERTIES:

Chemical Property	Description
Reactivity	Reacts with oxygen, acids, and halogens
Oxidation States	+3 (most stable), +2
Electronegativity	1.3 (Pauling scale)
Reaction with Water	Reacts slowly with water
Reaction with Acids	Dissolves in acids to form mendelevium salts
Reaction with Oxygen	Forms oxides similar to Md_2O_3
Flammability	Non-flammable, but mendelevium powder could ignite under certain conditions
Toxicity	Extremely toxic and radioactive, causing severe health risks
Biological Importance	No known biological role; harmful to living organisms due to its radioactivity
Isotopes	Mendelevium-258 (most stable), Mendelevium-257, Mendelevium-259 (used in research)

NOBELIUM

Hello Nobel fans! I'm Nobelium, named after Alfred Nobel. I'm synthetic and highly unstable, found mainly in nuclear research.

INTERESTING FACTS ABOUT NOBELIUM

❖Nobelium is named in honour of Alfred Nobel, the inventor of dynamite and the founder of the Nobel Prize.

❖Nobelium was first reported in 1957 by a group of scientists at the Nobel Institute of Physics in Sweden, but their discovery was disputed. The element was conclusively identified by a team at the Lawrence Berkeley National Laboratory in 1958.

❖Nobelium is a synthetic element, not found naturally on Earth, and is produced in particle accelerators by bombarding lighter elements with ions.

❖ Nobelium is highly radioactive, with its most stable isotope, Nobelium-259, having a half-life of 58 minutes.

❖ Nobelium is part of the actinide series of elements, which includes heavy, mostly radioactive elements.

❖ Due to its short half-life, nobelium exists only for brief periods, making it difficult to study extensively.

❖ Nobelium has no known practical applications due to its radioactivity and difficulty of producing it in significant quantities.

❖ Nobelium is unique among the actinides for having a stable +2 oxidation state, although the +3 oxidation state is also common.

❖ Nobelium is extremely rare, with only a few micrograms ever produced, making it one of the least studied elements.

❖ Nobelium shares chemical properties with other actinides, particularly californium and fermium.

❖ Nobelium is produced by bombarding curium with carbon ions in a particle accelerator, resulting in a fusion reaction that creates nobelium atoms.

❖ The element honours Alfred Nobel, who is not only known for his invention of dynamite but also for the prestigious prizes that bear his name.

PHYSICAL PROPERTIES:

Physical Property	Description
Atomic Symbol	No
Atomic Number	102
Atomic Mass	259 amu
State at Room Temp	Solid (metal)
Color	Unknown (likely silvery or metallic like other actinides)
Odor	Odourless
Density	Unknown (due to scarcity, physical properties are difficult to measure)
Melting Point	Estimated around 827°C
Boiling Point	Unknown
Solubility in Water	Insoluble in water
Conductivity	Likely a moderate conductor of electricity and heat, similar to other actinides
Abundance in Earth	Extremely rare, entirely synthetic

CHEMICAL PROPERTIES:

Chemical Property	Description
Reactivity	Reacts with oxygen, acids, and halogens
Oxidation States	+3 (most common), +2 (unusual for actinides but stable for nobelium)
Electronegativity	1.3 (Pauling scale)
Reaction with Water	Reacts slowly with water
Reaction with Acids	Dissolves in acids to form nobelium salts
Reaction with Oxygen	Forms oxides similar to No_2O_3
Flammability	Non-flammable, but could ignite under specific conditions
Toxicity	Extremely toxic and radioactive, posing severe health risks
Biological Importance	No known biological role; highly harmful to living organisms due to its radioactivity
Isotopes	Nobelium-259 (most stable), Nobelium-255, Nobelium-253 (used in nuclear research)

LAWRENCIUM

Hi research enthusiasts! I'm Lawrencium, the last of the actinides. You'll find me deep in the lab, helping scientists study the outer edges of the periodic table.

INTERESTING FACTS ABOUT LAWRENCIUM

❖Lawrencium was named in honour of Ernest O. Lawrence, the inventor of the cyclotron, a type of particle accelerator that has been critical in the discovery of many elements.

❖Lawrencium was first synthesised in 1961 by a team of scientists at the Lawrence Berkeley National Laboratory in California, including Albert Ghiorso, Torbjørn Sikkeland, and Almon Larsh.

❖Lawrencium is a synthetic element, does not occur naturally. It is produced by bombarding lighter elements in a particle accelerator.

❖ Lawrencium is highly radioactive, with its most stable isotope, Lawrencium-262, having a half-life of about 3.6 hours.

❖ Lawrencium is the last element in the actinide series, a group of heavy elements that are mostly radioactive.

❖ The relatively short half-life of lawrencium limits the time available for studying the element and its properties.

❖ Lawrencium typically exhibits a +3 oxidation state, similar to most actinides, although a +2 state has also been observed.

❖ Lawrencium is extremely rare, and only a few atoms have ever been produced, making it one of the least studied elements.

❖ Lawrencium is synthesised by bombarding californium-252 with boron ions, producing lawrencium atoms through nuclear fusion.

❖ The element's name honours Ernest Lawrence, who played a key role in the advancement of nuclear physics and the development of cyclotron technology.

❖ The discovery of lawrencium was initially claimed by both American and Soviet scientists, but the International Union of Pure and Applied Chemistry (IUPAC) credited the American team with the official discovery.

PHYSICAL PROPERTIES:

Physical Property	Description
Atomic Symbol	Lr
Atomic Number	103
Atomic Mass	262 amu
State at Room Temp	Solid (metal)
Color	Unknown (likely silvery or metallic like other actinides)
Odor	Odourless
Density	Unknown (due to scarcity, physical properties are difficult to measure)
Melting Point	Estimated around 1,627°C
Boiling Point	Unknown
Solubility in Water	Insoluble in water
Conductivity	Likely a moderate conductor of electricity and heat, similar to other actinides
Abundance in Earth	Extremely rare, entirely synthetic

CHEMICAL PROPERTIES:

Chemical Property	Description
Reactivity	Reacts with oxygen, acids, and halogens
Oxidation States	+3 (most stable), +2 (less common but observed)
Electronegativity	1.3 (Pauling scale)
Reaction with Water	Reacts slowly with water
Reaction with Acids	Dissolves in acids to form lawrencium salts
Reaction with Oxygen	Forms oxides similar to Lr_2O_3
Flammability	Non-flammable, but could ignite under specific conditions
Toxicity	Extremely toxic and radioactive, posing severe health risks
Biological Importance	No known biological role; hazardous to living organisms due to its radioactivity
Isotopes	Lawrencium-262 (most stable), Lawrencium-261, Lawrencium-260 (used in nuclear research)

RUTHERFORDIUM

Greetings element discoverers! I'm Rutherfordium, named after the father of nuclear physics. I'm part of the mysterious world of superheavy elements.

INTERESTING FACTS ABOUT RUTHERFORDIUM

❖Rutherfordium is named in honour of Ernest Rutherford, the father of nuclear physics, who discovered the concept of radioactive half-lives and the structure of the atom.

❖Rutherfordium was first synthesised in 1969 by a team of scientists at the Joint Institute for Nuclear Research (JINR) in Dubna, Russia, and independently confirmed by scientists at the Lawrence Berkeley National Laboratory in California.

❖Rutherfordium is the first element in the transactinide series, elements that are heavier than the actinides.

❖ Rutherfordium is a synthetic element that does not occur naturally and must be created in particle accelerators by bombarding lighter elements with ions.

❖ Rutherfordium is highly radioactive, with its most stable isotope, Rutherfordium-267, having a half-life of about 1.3 hours.

❖ Several isotopes of rutherfordium have been synthesised, with Rutherfordium-267 being the most stable, though all isotopes are highly unstable.

❖ The discovery of rutherfordium was disputed between Soviet and American scientists, with both claiming to have synthesised it first. The International Union of Pure and Applied Chemistry (IUPAC) credited both groups with the discovery but named the element after Rutherford.

❖ Due to its radioactivity and the difficulty in producing it, rutherfordium has no known commercial or industrial applications and is used only for scientific research.

❖ Only a few atoms of rutherfordium have ever been produced, making it one of the least studied elements.

PHYSICAL PROPERTIES:

Physical Property	Description
Atomic Symbol	Rf
Atomic Number	104
Atomic Mass	267 amu
State at Room Temp	Solid (metal)
Color	Unknown (likely silvery or metallic like other transition metals)
Odor	Odourless
Density	Estimated around 23 g/cm³
Melting Point	Unknown
Boiling Point	Unknown
Solubility in Water	Insoluble in water
Conductivity	Likely a conductor of electricity and heat, similar to other group 4 elements
Abundance in Earth	Extremely rare, entirely synthetic

CHEMICAL PROPERTIES:

Chemical Property	Description
Reactivity	Expected to react with oxygen, acids, and halogens
Oxidation States	+4 (most common), possibly +3
Electronegativity	Estimated to be 1.3 (Pauling scale)
Reaction with Water	Expected to react with water slowly, forming oxides
Reaction with Acids	Expected to dissolve in acids to form rutherfordium salts
Reaction with Oxygen	Expected to form oxides, similar to RfO_2
Flammability	Non-flammable
Toxicity	Highly toxic and radioactive, posing severe health risks
Biological Importance	No known biological role; harmful to living organisms due to its radioactivity
Isotopes	Rutherfordium-267 (most stable), Rutherfordium-263, Rutherfordium-261 (used in research)

DUBNIUM

Hello researchers! I'm Dubnium, a synthetic element with a very short lifespan. You'll find me in labs helping scientists explore the unknown.

INTERESTING FACTS ABOUT DUBNIUM

❖Dubnium is named after the town of Dubna in Russia, home to the Joint Institute for Nuclear Research (JINR), where it was first synthesised.

❖Dubnium was first synthesised in 1967 by scientists at JINR in Dubna, Russia, though American scientists at the Lawrence Berkeley National Laboratory in California also claimed its discovery in 1970.

❖The discovery of dubnium was hotly contested between Soviet and American scientists, with both claiming to have synthesised the element first. The naming of the element was also debated until IUPAC officially named it after Dubna in 1997.

❖ Dubnium is a highly radioactive element, with no stable isotopes. Its most stable isotope, Dubnium-268, has a half-life of approximately 32 hours.

❖ Dubnium is a transactinide element, a group of heavy, mostly synthetic elements that come after actinides in the periodic table.

❖ Dubnium does not occur naturally and is produced artificially in particle accelerators through the collision of lighter elements.

❖ Because of its radioactivity and short half-life, dubnium has no practical applications and is used only in scientific research.

❖ Dubnium belongs to group 5 in the periodic table, alongside vanadium, niobium, and tantalum, which suggests it might have similar chemical properties, though these are difficult to study due to its radioactivity.

❖ Dubnium is expected to exhibit oxidation states of +5, which is common for other group 5 elements, but lower oxidation states like +3 have also been predicted.

❖ Dubnium is produced by bombarding californium with nitrogen ions or by using other heavy ion fusion reactions.

PHYSICAL PROPERTIES:

Physical Property	Description
Atomic Symbol	Db
Atomic Number	105
Atomic Mass	268 amu
State at Room Temp	Solid (metal)
Color	Unknown (likely silvery or metallic like other transition metals)
Odor	Odourless
Density	Estimated around 29 g/cm^3
Melting Point	Unknown
Boiling Point	Unknown
Solubility in Water	Insoluble in water
Conductivity	Likely a conductor of electricity and heat, similar to other group 5 elements
Abundance in Earth	Extremely rare, entirely synthetic

CHEMICAL PROPERTIES:

Chemical Property	Description
Reactivity	Expected to react with oxygen, acids, and halogens
Oxidation States	+5 (most common), possibly +3
Electronegativity	Estimated to be around 1.3 (Pauling scale)
Reaction with Water	Expected to react slowly with water, forming oxides
Reaction with Acids	Expected to dissolve in acids, forming dubnium salts
Reaction with Oxygen	Expected to form oxides similar to Db_2O_5
Flammability	Non-flammable
Toxicity	Extremely toxic and radioactive, posing severe health risks
Biological Importance	No known biological role; harmful to living organisms due to its radioactivity
Isotopes	Dubnium-268 (most stable), Dubnium-270, Dubnium-262 (used in research)

SEABORGIUM

Hi scientists! I'm Seaborgium, named after Glenn Seaborg, a pioneer of actinide research. I'm part of the superheavy elements that push the limits of chemistry.

INTERESTING FACTS ABOUT SEABORGIUM

❖Seaborgium is named in honour of American chemist Glenn T. Seaborg, who contributed to the discovery of many transuranium elements and the development of the actinide concept.

❖Seaborgium was first synthesised in 1974 by a team of scientists at the Joint Institute for Nuclear Research (JINR) in Dubna, Russia, and the Lawrence Berkeley National Laboratory in California, USA.

❖Seaborgium belongs to the transactinide elements, a group of synthetic elements that are heavier than the actinides and follow uranium in the periodic table.

❖ The discovery of seaborgium was the subject of controversy, as both the Russian and American teams claimed to have synthesised the element first. However, IUPAC eventually credited both groups.

❖ Seaborgium is highly radioactive, with its most stable isotope, Seaborgium-271, having a half-life of about 2.4 minutes.

❖ Like other transactinide elements, seaborgium does not occur naturally and must be synthesised in a laboratory using particle accelerators.

❖ Due to its radioactivity and extremely short half-life, seaborgium has no known commercial or industrial applications and is only used in scientific research.

❖ Seaborgium is part of group 6 in the periodic table, along with chromium, molybdenum, and tungsten, suggesting it may exhibit similar chemical properties, though these are difficult to study.

❖ Seaborgium is produced by bombarding californium or uranium targets with lighter elements such as oxygen or neon in a particle accelerator, causing nuclear fusion.

❖ Seaborgium was the first element to be named after a person while they were still alive-Glenn T. Seaborg received this honour in 1997.

PHYSICAL PROPERTIES:

Physical Property	Description
Atomic Symbol	Sg
Atomic Number	106
Atomic Mass	271 amu
State at Room Temp	Solid (metal)
Color	Unknown (likely silvery or metallic like other transition metals)
Odor	Odourless
Density	Estimated around 35 g/cm³
Melting Point	Unknown
Boiling Point	Unknown
Solubility in Water	Insoluble in water
Conductivity	Likely a conductor of electricity and heat, similar to other group 6 elements
Abundance in Earth	Extremely rare, entirely synthetic

CHEMICAL PROPERTIES:

Chemical Property	Description
Reactivity	Expected to react with oxygen, acids, and halogens
Oxidation States	+6 (most common), possibly +4
Electronegativity	Estimated to be around 1.3 (Pauling scale)
Reaction with Water	Expected to react slowly with water, forming oxides
Reaction with Acids	Expected to dissolve in acids, forming seaborgium salts
Reaction with Oxygen	Expected to form oxides, similar to SgO_3
Flammability	Non-flammable
Toxicity	Highly toxic and radioactive, posing severe health risks
Biological Importance	No known biological role; harmful to living organisms due to its radioactivity
Isotopes	Seaborgium-271 (most stable), Seaborgium-270, Seaborgium-269 (used in research)

BOHRIUM

Hello physicists! I'm Bohrium, a synthetic and radioactive element. While my existence is short, I help researchers understand the properties of heavy elements.

INTERESTING FACTS ABOUT BOHRIUM

❖Bohrium is named in honour of the famous Danish physicist Niels Bohr, who made significant contributions to our understanding of atomic structure and quantum theory.

❖Bohrium was first synthesised in 1981 by a team of German scientists led by Peter Armbruster and Gottfried Münzenberg at the Gesellschaft für Schwerionenforschung (GSI) in Darmstadt, Germany.

❖Bohrium belongs to the transactinide elements, which are synthetic, radioactive elements heavier than uranium.

❖ Bohrium is highly radioactive, with its most stable isotope, Bohrium-270, having a half-life of about 61 seconds.

❖ It does not occur naturally, it is synthesised artificially in particle accelerators by bombarding lighter elements with heavier ones.

❖ Bohrium has no known practical applications due to its short half-life and the difficulty in producing it. It is mainly used for scientific research purposes.

❖ Bohrium is produced by bombarding bismuth (element 83) with chromium ions (element 24) in a particle accelerator, creating a fusion reaction that results in bohrium atoms.

❖ Bohrium is expected to exhibit a +7 oxidation state, similar to other group 7 elements, but its chemistry is not well understood.

❖ Due to the extremely short half-life of bohrium isotopes, it is very difficult to study its physical and chemical properties in detail.

❖ Only a few atoms of bohrium have ever been created, making it one of the rarest and least studied elements.

❖ Bohrium's synthesis and study contribute to our understanding of nuclear physics and the stability of heavy elements at the end of the periodic table.

PHYSICAL PROPERTIES:

Physical Property	Description
Atomic Symbol	Bh
Atomic Number	107
Atomic Mass	270 amu
State at Room Temp	Solid (metal)
Color	Unknown (likely silvery or metallic like other transition metals)
Odor	Odourless
Density	Estimated around 37 g/cm^3
Melting Point	Unknown
Boiling Point	Unknown
Solubility in Water	Insoluble in water
Conductivity	Likely a conductor of electricity and heat, similar to other group 7 elements
Abundance in Earth	Extremely rare, entirely synthetic

CHEMICAL PROPERTIES:

Chemical Property	Description
Reactivity	Expected to react with oxygen, acids, and halogens
Oxidation States	+7 (most common), possibly lower states such as +5 or +3
Electronegativity	Estimated to be around 1.3 (Pauling scale)
Reaction with Water	Expected to react slowly with water, forming oxides
Reaction with Acids	Expected to dissolve in acids, forming bohrium salts
Reaction with Oxygen	Expected to form oxides, similar to Bh_2O_7 or BhO_3
Flammability	Non-flammable
Toxicity	Highly toxic and radioactive, posing severe health risks
Biological Importance	No known biological role; harmful to living organisms due to its radioactivity
Isotopes	Bohrium-270 (most stable), Bohrium-272, Bohrium-264 (used in research)

HASSIUM

Hi atomic explorers! I'm Hassium, a highly radioactive and short-lived element. I'm named after the German state of Hesse, where I was first synthesised.

INTERESTING FACTS ABOUT HASSIUM

❖Hassium is named after the German state of Hesse, where the element was first synthesized at the Gesellschaft für Schwerionenforschung (GSI) in Darmstadt.

❖Hassium was first synthesised in 1984 by a team of German scientists led by Peter Armbruster & Gottfried Münzenberg at GSI.

❖Hassium belongs to the transactinide elements, a group of synthetic elements that are heavier than uranium and produced in laboratories.

❖Hassium is highly radioactive, with its most stable isotope, Hassium-277, having a half-life of approximately 11 minutes.

❖ Hassium does not occur naturally and must be synthesised in a particle accelerator by bombarding heavier elements with lighter ones.

❖ Due to its radioactivity and short half-life, hassium has no known practical applications and is used only in scientific research.

❖ Hassium is part of group 8 in the periodic table, alongside iron, ruthenium, and osmium, suggesting it may have similar chemical properties.

❖ Hassium is expected to exhibit a +8 oxidation state, similar to osmium and ruthenium, although this has not been extensively confirmed.

❖ Hassium is produced by bombarding lead with iron ions in a particle accelerator, causing a nuclear fusion reaction.

❖ Only a few atoms of hassium have been synthesised, making it an extremely rare and difficult-to-study element.

❖ Theoretical studies and limited experimental data suggest that hassium may form a compound called hassium tetroxide (HsO_4), similar to osmium tetroxide.

PHYSICAL PROPERTIES:

Physical Property	Description
Atomic Symbol	Hs
Atomic Number	108
Atomic Mass	277 amu
State at Room Temp	Solid (metal)
Color	Unknown (likely silvery or metallic like other transition metals)
Odor	Odourless
Density	Estimated around 40 g/cm³
Melting Point	Unknown
Boiling Point	Unknown
Solubility in Water	Insoluble in water
Conductivity	Likely a conductor of electricity and heat, similar to other group 8 elements
Abundance in Earth	Extremely rare, entirely synthetic

CHEMICAL PROPERTIES:

Chemical Property	Description
Reactivity	Expected to react with oxygen, acids, and halogens
Oxidation States	+8 (most common), possibly lower states like +6 and +4
Electronegativity	Estimated to be around 1.3 (Pauling scale)
Reaction with Water	Expected to react slowly with water, forming oxides
Reaction with Acids	Expected to dissolve in acids, forming hassium salts
Reaction with Oxygen	Expected to form oxides, similar to HsO_4 (hassium tetroxide)
Flammability	Non-flammable
Toxicity	Extremely toxic and radioactive, posing severe health risks
Biological Importance	No known biological role; harmful to living organisms due to its radioactivity
Isotopes	Hassium-277 (most stable), Hassium-270, Hassium-269 (used in research)

MEITNERIUM

Hello trailblazers! I'm Meitnerium, named after Lise Meitner, a physicist who helped discover nuclear fission. I'm one of the newest elements on the block.

INTERESTING FACTS ABOUT MEITNERIUM

❖Meitnerium is named in honour of Austrian-Swedish physicist Lise Meitner, who contributed significantly to the discovery of nuclear fission alongside Otto Hahn.

❖Meitnerium was first synthesized in 1982 by a team of German scientists at the Gesellschaft für Schwerionenforschung (GSI) in Darmstadt, led by Peter Armbruster and Gottfried Münzenberg.

❖Meitnerium is one of the transactinide elements, which are synthetic and highly radioactive, existing only through laboratory synthesis.

❖ Meitnerium is a highly radioactive element. Its most stable isotope, Meitnerium-278, has a half-life of about 7.6 seconds, making it extremely difficult to study.

❖ Like other transactinide elements, meitnerium does not occur naturally and is only created in particle accelerators by bombarding heavy elements.

❖ Due to its short half-life and instability, meitnerium has no known practical or commercial uses. It is primarily studied for scientific research.

❖ Meitnerium belongs to group 9 in the periodic table, along with cobalt, rhodium, and iridium. Its chemical properties are expected to resemble those of iridium.

❖ Meitnerium is expected to have a +3 oxidation state, similar to its group counterparts, although its chemistry remains largely unexplored.

❖ Meitnerium is produced by bombarding bismuth (element 83) with iron ions (element 26) in a particle accelerator, creating a fusion reaction that produces meitnerium atoms.

❖ Only a small number of meitnerium atoms have been synthesised.

PHYSICAL PROPERTIES:

Physical Property	Description
Atomic Symbol	Mt
Atomic Number	109
Atomic Mass	278 amu
State at Room Temp	Solid (metal)
Color	Unknown (likely silvery or metallic like other transition metals)
Odor	Odourless
Density	Estimated around 37.4 g/cm³
Melting Point	Unknown
Boiling Point	Unknown
Solubility in Water	Insoluble in water
Conductivity	Likely a conductor of electricity and heat, similar to other group 9 elements
Abundance in Earth	Extremely rare, entirely synthetic

CHEMICAL PROPERTIES:

Chemical Property	Description
Reactivity	Expected to react with oxygen, acids, and halogens
Oxidation States	+3 (most common), potentially lower oxidation states like +1 or +2
Electronegativity	Estimated to be around 1.3 (Pauling scale)
Reaction with Water	Expected to react slowly with water, forming oxides
Reaction with Acids	Expected to dissolve in acids, forming meitnerium salts
Reaction with Oxygen	Expected to form oxides, similar to Mt_2O_3
Flammability	Non-flammable
Toxicity	Extremely toxic and radioactive, posing severe health risks
Biological Importance	No known biological role; harmful to living organisms due to its radioactivity
Isotopes	Meitnerium-278 (most stable), Meitnerium-276, Meitnerium-274 (used in research)

DARMSTADTIUM

Hi particle physicists! I'm Darmstadtium, named after the German city where I was discovered. I'm a superheavy element, found only in particle accelerators.

INTERESTING FACTS ABOUT DARMSTADTIUM

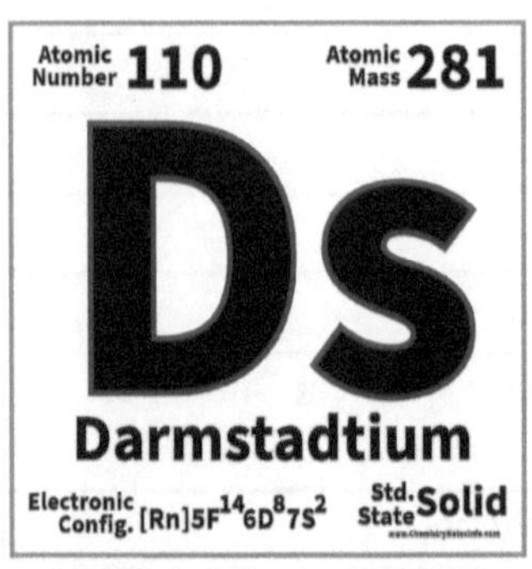

❖Darmstadtium is named after the city of Darmstadt in Germany, where it was first synthesized at the Gesellschaft für Schwerionenforschung (GSI) in 1994.

❖Darmstadtium was discovered in 1994 by a team of scientists led by Peter Armbruster and Gottfried Münzenberg at GSI, Darmstadt.

❖Darmstadtium is part of the transactinide elements, which are superheavy, synthetic elements with atomic no. greater than 100.

❖Darmstadtium is highly radioactive, and its most stable isotope, Darmstadtium-281, has a half-life of around 20 seconds.

❖ Darmstadtium does not occur naturally and is synthesised in laboratories through particle accelerator experiments.

❖ Due to its very short half-life, darmstadtium has no known practical applications outside of scientific research.

❖ Darmstadtium is part of group 10 in the periodic table, along with nickel, palladium, and platinum. Its chemical properties are expected to resemble those of platinum.

❖ Darmstadtium is synthesised by bombarding a lead (Pb) target with nickel (Ni) ions, causing a fusion reaction that creates darmstadtium atoms.

❖ Only a few atoms of darmstadtium have ever been produced, making it one of the most challenging elements to study.

❖ Due to the difficulty in studying its chemistry, very little is known about darmstadtium's chemical behaviour, although it is predicted to be a noble metal.

❖ Most research on darmstadtium focuses on understanding stability of superheavy elements and their potential to form stable isotopes.

PHYSICAL PROPERTIES:

Physical Property	Description
Atomic Symbol	Ds
Atomic Number	110
Atomic Mass	281 amu
State at Room Temp	Solid (metal)
Color	Unknown (likely silvery or metallic like other transition metals)
Odor	Odourless
Density	Estimated around 34.8 g/cm³
Melting Point	Unknown
Boiling Point	Unknown
Solubility in Water	Insoluble in water
Conductivity	Likely a conductor of electricity and heat, similar to other group 10 elements
Abundance in Earth	Extremely rare, entirely synthetic

CHEMICAL PROPERTIES:

Chemical Property	Description
Reactivity	Expected to be chemically inert, similar to platinum and other noble metals
Oxidation States	Predicted to have +4, +2, and possibly +6 oxidation states
Electronegativity	Estimated to be around 1.3 (Pauling scale)
Reaction with Water	Likely unreactive with water due to its expected noble character
Reaction with Acids	Likely resistant to acids, similar to platinum
Reaction with Oxygen	Predicted to be resistant to oxidation, possibly forming oxides at high temperatures
Flammability	Non-flammable
Toxicity	Extremely toxic and radioactive, posing severe health risks
Biological Importance	No known biological role; harmful to living organisms due to its radioactivity
Isotopes	Darmstadtium-281 (most stable), Darmstadtium-280, Darmstadtium-279 (used in research)

ROENTGENIUM

Hello X-ray fans! I'm Roentgenium, named after Wilhelm Röntgen, the discoverer of X-rays. My life is brief in the lab, but I honour a legacy of revolutionary discoveries in physics.

INTERESTING FACTS ABOUT ROENTGENIUM

❖Roentgenium is named in honour of Wilhelm Conrad Roentgen, the German physicist who discovered X-rays in 1895.

❖Roentgenium was first synthesised in 1994 by a team of scientists at Gesellschaft für Schwerionenforschung (GSI) in Darmstadt, Germany.

❖Roentgenium is a highly radioactive element, and its most stable isotope, Roentgenium-282, has a half-life of about 2.1 minutes.

❖Roentgenium does not occur naturally. It is a synthetic element that is produced in laboratories using particle accelerators.

❖ Due to its short half-life and instability, roentgenium has no known practical applications and is used only in scientific research.

❖ Roentgenium is a member of the transactinide elements, which are superheavy and exist only through artificial synthesis.

❖ Roentgenium is part of group 11 in the periodic table, along with copper, silver, and gold. Its chemical properties are predicted to be similar to gold.

❖ Roentgenium is produced by bombarding bismuth (element 83) with nickel ions (element 28) in a particle accelerator, resulting in the fusion of these nuclei to form roentgenium.

❖ Only a handful of atoms of roentgenium have ever been synthesised, making it one of the rarest elements known.

❖ Roentgenium is expected to behave like a noble metal, with chemical properties resembling those of gold and silver, although it has been difficult to study experimentally.

❖ Roentgenium's placement near the "island of stability" in the periodic table makes it of interest to scientists studying superheavy elements and their potential for more stable isotopes.

PHYSICAL PROPERTIES:

Physical Property	Description
Atomic Symbol	Rg
Atomic Number	111
Atomic Mass	282 amu
State at Room Temp	Solid (metal)
Color	Unknown (likely metallic or gold-like based on its predicted properties)
Odor	Odourless
Density	Estimated around 28.7 g/cm³
Melting Point	Unknown
Boiling Point	Unknown
Solubility in Water	Insoluble in water
Conductivity	Likely a good conductor of electricity and heat, similar to other group 11 elements
Abundance in Earth	Extremely rare, entirely synthetic

CHEMICAL PROPERTIES:

Chemical Property	Description
Reactivity	Expected to behave like gold, with resistance to corrosion and reaction with most chemicals
Oxidation States	+1 (most common), possibly +3, although experimental data is lacking
Electronegativity	Estimated to be around 1.9 (Pauling scale)
Reaction with Water	Predicted to be unreactive with water, similar to gold
Reaction with Acids	Likely resistant to most acids, similar to gold's inertness
Reaction with Oxygen	Likely forms oxides at high temperatures, but generally resistant to oxidation at room temperature
Flammability	Non-flammable
Toxicity	Extremely toxic and radioactive, posing severe health risks
Biological Importance	No known biological role; harmful to living organisms due to its radioactivity
Isotopes	Roentgenium-282 (most stable), Roentgenium-281, Roentgenium-280 (used in research)

COPERNICIUM

Hi stargazers! I'm Copernicium, named after Nicolaus Copernicus, the astronomer who reshaped our view of the universe. I'm a superheavy element, created in particle accelerators and studied for my fleeting yet fascinating properties.

INTERESTING FACTS ABOUT COPERNICIUM

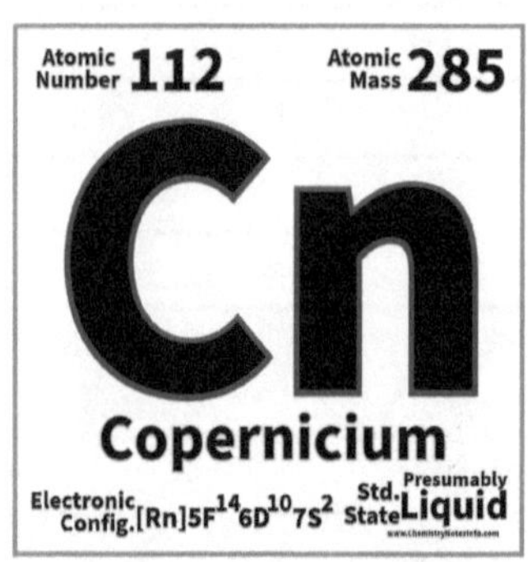

❖Copernicium is named in honour of the astronomer Nicolaus Copernicus, who formulated heliocentric model of the solar system.

❖Copernicium was first synthesised on February 9, 1996, by a team of scientists at the Gesellschaft für Schwerionenforschung (GSI) in Darmstadt, Germany.

❖Copernicium is a highly radioactive element. Its most stable isotope, Copernicium-285, has a half-life of about 29 seconds.

❖Copernicium does not occur naturally; it is a synthetic element produced in laboratories.

❖ Copernicium is a transactinide element, which is a superheavy element with an atomic number above 100.

❖ Copernicium is part of group 12 in the periodic table, along with zinc, cadmium, and mercury. It is expected to behave like a volatile metal similar to mercury.

❖ Copernicium is produced by bombarding lead (element 82) with zinc (element 30) ions in a particle accelerator, leading to the fusion of these elements and the creation of copernicium atoms.

❖ Only a few atoms of copernicium have ever been synthesised, making it one of the rarest elements known.

❖ Copernicium is predicted to be a volatile metal, like mercury, meaning it could exist as a liquid at or near room temperature.

❖ Due to its short half-life and the difficulty of working with superheavy elements, the chemical properties of copernicium remain largely unexplored.

❖ Although discovered in 1996, copernicium received its official name from the IUPAC in 2010.

PHYSICAL PROPERTIES:

Physical Property	Description
Atomic Symbol	Cn
Atomic Number	112
Atomic Mass	285 amu
State at Room Temp	Solid or possibly liquid (predicted to behave like mercury)
Color	Unknown (likely metallic or silvery in appearance)
Odor	Odourless
Density	Estimated around 14 g/cm³ (similar to mercury)
Melting Point	Unknown (predicted to be low like mercury)
Boiling Point	Unknown
Solubility in Water	Insoluble in water
Conductivity	Expected to be a good conductor of electricity and heat, similar to other group 12 elements
Abundance in Earth	Extremely rare, entirely synthetic

CHEMICAL PROPERTIES:

Chemical Property	Description
Reactivity	Expected to be less reactive than mercury, possibly forming weak interactions with other elements
Oxidation States	Predicted to have a +2 oxidation state, similar to mercury
Electronegativity	Estimated to be around 1.6 (Pauling scale)
Reaction with Water	Expected to be unreactive with water, like mercury
Reaction with Acids	Likely resistant to most acids, but may form salts in strong acidic conditions
Reaction with Oxygen	Expected to resist oxidation under normal conditions, forming oxides only at high temperatures
Flammability	Non-flammable
Toxicity	Extremely toxic and radioactive, posing severe health risks
Biological Importance	No known biological role; harmful to living organisms due to its radioactivity
Isotopes	Copernicium-285 (most stable), Copernicium-283, Copernicium-284 (used in research)

NIHONIUM

Hello innovators! I'm Nihonium, the first element discovered in Japan. While I exist for only a short time in the lab, my discovery is a milestone in advancing heavy element research.

INTERESTING FACTS ABOUT NIHONIUM

❖Nihonium is named after "Nihon," the Japanese word for Japan, as the element was first discovered by a Japanese research team.

❖Nihonium was first synthesised on July 23, 2003, by a team of scientists at the RIKEN Institute in Japan, led by Kosuke Morita.

❖Nihonium is a highly radioactive element, and its most stable isotope, Nihonium-286, has a half-life of about 10 seconds.

❖Nihonium is historically significant as the first element discovered in Asia, specifically by a Japanese team.

❖ Like other superheavy elements, Nihonium is part of the transactinide series, which are elements with atomic numbers greater than 100.

❖ Nihonium belongs to group 13 in the periodic table, along with boron, aluminium, gallium, indium, and thallium. Its chemical properties are expected to be similar to thallium.

❖ Nihonium is produced by bombarding a bismuth (element 83) target with zinc (element 30) ions in a particle accelerator, leading to a fusion reaction that creates nihonium atoms.

❖ Only a handful of nihonium atoms have been synthesised, making it extremely difficult to study.

❖ The discovery of nihonium completed group 13 of the periodic table, making it chemically significant despite its fleeting existence.

❖ Nihonium is expected to be a metal and might behave similarly to other group 13 metals, although its short half-life limits the study of its chemical properties.

❖ Due to its very short half-life and high radioactivity, nihonium has no practical applications and is used only for research purposes.

❖ Although discovered in 2003, It was officially named by International Union of Pure and Applied Chemistry (IUPAC) in 2016.

PHYSICAL PROPERTIES:

Physical Property	Description
Atomic Symbol	Nh
Atomic Number	113
Atomic Mass	286 amu
State at Room Temp	Solid (metal, predicted)
Color	Unknown (likely metallic or silvery like other group 13 metals)
Odor	Odorless
Density	Estimated around 16 g/cm³
Melting Point	Unknown
Boiling Point	Unknown
Solubility in Water	Insoluble in water
Conductivity	Expected to be a conductor of electricity and heat, like other group 13 metals
Abundance in Earth	Extremely rare, entirely synthetic

CHEMICAL PROPERTIES:

Chemical Property	Description
Reactivity	Predicted to behave like thallium, with a tendency to form +1 and +3 oxidation states
Oxidation States	+1 and +3 (predicted, based on similarity to thallium)
Electronegativity	Estimated to be around 1.6 (Pauling scale)
Reaction with Water	Likely to be unreactive with water, forming no stable compounds due to its short half-life
Reaction with Acids	May form simple compounds in acidic environments, but data is lacking
Reaction with Oxygen	Likely forms oxides, but would decompose rapidly due to its radioactivity
Flammability	Non-flammable
Toxicity	Extremely toxic and radioactive, posing severe health risks
Biological Importance	No known biological role; harmful to living organisms due to its radioactivity
Isotopes	Nihonium-286 (most stable), Nihonium-285, Nihonium-284 (used in research)

FLEROVIUM

Hi element enthusiasts! I'm Flerovium, named after the Flerov Laboratory of Nuclear Reactions. I'm part of the cutting-edge research into superheavy elements, where scientists are exploring the limits of stability.

INTERESTING FACTS ABOUT FLEROVIUM

❖ Flerovium is named after the Flerov Laboratory of Nuclear Reactions in Dubna, Russia, which was itself named after the Soviet physicist Georgy Flerov.

❖ Flerovium was first synthesised on December 30, 1998, by a team of scientists at the Joint Institute for Nuclear Research (JINR) in Dubna, Russia, in collaboration with the Lawrence Livermore National Laboratory in the USA.

❖ Flerovium is a highly radioactive element, and its most stable isotope, Flerovium-289, has a half-life of around 2.6 seconds.

❖ Like other superheavy elements, Flerovium does not occur naturally. It is a synthetic element produced in laboratories.

❖ It is a transactinide element, which means it belongs to a group of superheavy elements with atomic numbers greater than 100.

❖ Flerovium is part of group 14 in the periodic table, alongside carbon, silicon, germanium, tin, and lead. Its properties are predicted to be similar to lead.

❖ Flerovium is created by bombarding plutonium-244 with calcium-48 ions in a particle accelerator, leading to the fusion of these elements.

❖ Some predictions suggest that Flerovium may behave like a noble gas or noble metal, showing limited chemical reactivity.

❖ While Flerovium's properties are largely unknown, it is expected to behave like a very heavy metal, with metallic characteristics.

❖ Only a small number of Flerovium atoms have been synthesised, making it one of the rarest elements.

❖ Flerovium was officially named by IUPAC in May 2012, over a decade after its discovery.

PHYSICAL PROPERTIES:

Physical Property	Description
Atomic Symbol	Fl
Atomic Number	114
Atomic Mass	289 amu
State at Room Temp	Solid (predicted to be a metal)
Color	Unknown (likely metallic or silvery in appearance)
Odor	Odorless
Density	Estimated around 14 g/cm^3
Melting Point	Unknown
Boiling Point	Unknown
Solubility in Water	Insoluble in water
Conductivity	Expected to be a conductor of electricity and heat, similar to lead
Abundance in Earth	Extremely rare, entirely synthetic

CHEMICAL PROPERTIES:

Chemical Property	Description
Reactivity	Expected to be less reactive than lead, potentially inert like noble gases or noble metals
Oxidation States	Predicted to have +2 and +4 oxidation states, similar to lead
Electronegativity	Estimated to be around 1.3 (Pauling scale)
Reaction with Water	Likely to be unreactive with water, similar to lead
Reaction with Acids	Expected to form some compounds in acidic conditions, but likely to be fairly inert
Reaction with Oxygen	May form oxides, but would decompose rapidly due to its radioactivity
Flammability	Non-flammable
Toxicity	Extremely toxic and radioactive, posing severe health risks
Biological Importance	No known biological role; harmful to living organisms due to its radioactivity
Isotopes	Flerovium-289 (most stable), Flerovium-288, Flerovium-287 (used in research)

MOSCOVIUM

Hello element hunters! I'm Moscovium, named after Moscow, Russia. I'm a synthetic, superheavy element, created in labs, and though I'm unstable, my discovery opens new doors in the quest for ultra-heavy elements.

INTERESTING FACTS ABOUT MOSCOVIUM

❖Moscovium is named in honour of the Moscow region, reflecting the contributions of Russian scientists at the Joint Institute for Nuclear Research (JINR) in Dubna.

❖Moscovium was first synthesised on July 2, 2003, by a team of Russian and American scientists at JINR in collaboration with Lawrence Livermore National Laboratory.

❖Moscovium is a highly radioactive element, and its most stable isotope, Moscovium-290, has a half-life of around 0.65 seconds.

❖ Moscovium is a transactinide element, belonging to the group of superheavy elements with atomic numbers above 100.

❖ Moscovium is part of group 15 in the periodic table, along with nitrogen, phosphorus, arsenic, antimony, and bismuth. Its properties are expected to be similar to bismuth.

❖ Moscovium is produced by bombarding americium-243 with calcium-48 ions in a particle accelerator, resulting in the fusion of the elements.

❖ Only a few atoms of moscovium have ever been synthesised, making it extremely rare and difficult to study.

❖ Moscovium is expected to behave like a metal, though its chemical and physical properties remain largely unexplored due to its short half-life.

❖ Moscovium was officially named by the International Union of Pure and Applied Chemistry (IUPAC) in November 2016.

❖ Due to short half-life & high radioactivity, moscovium has no known practical applications & is used only for scientific research.

❖ Moscovium is likely a post-transition metal, with properties similar to other heavier elements in group 15, such as bismuth.

PHYSICAL PROPERTIES:

Physical Property	Description
Atomic Symbol	Mc
Atomic Number	115
Atomic Mass	290 amu
State at Room Temp	Solid (metal, predicted)
Color	Unknown (likely metallic or silvery like bismuth)
Odor	Odourless
Density	Estimated around 13.5 g/cm³
Melting Point	Unknown
Boiling Point	Unknown
Solubility in Water	Insoluble in water
Conductivity	Expected to be a good conductor of electricity and heat, like other metals
Abundance in Earth	Extremely rare, entirely synthetic

CHEMICAL PROPERTIES:

Chemical Property	Description
Reactivity	Predicted to behave similarly to bismuth, forming compounds with a +1 or +3 oxidation state
Oxidation States	+1 and +3 (predicted)
Electronegativity	Estimated to be around 1.3 (Pauling scale)
Reaction with Water	Likely to be unreactive with water
Reaction with Acids	Expected to form salts in strong acidic environments
Reaction with Oxygen	May form oxides, though it would decompose rapidly due to its short half-life
Flammability	Non-flammable
Toxicity	Extremely toxic and radioactive, posing severe health risks
Biological Importance	No known biological role; harmful to living organisms due to its radioactivity
Isotopes	Moscovium-290 (most stable), Moscovium-289, Moscovium-288 (used in research)

LIVERMORIUM

Hi lab researchers! I'm Livermorium, named after the Lawrence Livermore National Laboratory. I'm part of the heavy-element frontier, synthesised in labs, where scientists study me to understand the forces that hold atoms together.

INTERESTING FACTS ABOUT LIVERMORIUM

❖Livermorium is named after the Lawrence Livermore National Laboratory (LLNL) in California, in recognition of the lab's collaboration with Russian scientists in the discovery of several superheavy elements.

❖Livermorium was first synthesised on July 19, 2000, by a team of scientists at the Joint Institute for Nuclear Research (JINR) in Dubna, Russia, in collaboration with LLNL.

❖Livermorium is a highly radioactive element, and its most stable isotope, Livermorium-293, has a half-life of about 60 milliseconds.

❖ Livermorium is not found in nature and can only be produced in a laboratory setting through nuclear reactions.

❖ Livermorium belongs to group 16 in the periodic table, along with oxygen, sulphur, selenium, tellurium, and polonium. Its properties are predicted to resemble those of polonium.

❖ Livermorium is part of the transactinide series, which includes superheavy elements with atomic numbers greater than 100.

❖ Livermorium is produced by bombarding curium-248 with calcium-48 ions in a particle accelerator, leading to the fusion of these elements.

❖ Only a few atoms of livermorium have been synthesised, making it extremely difficult to study its properties.

❖ Livermorium is expected to behave as a metal, although its physical and chemical properties remain largely speculative due to its short half-life.

❖ Livermorium was officially named by International Union of Pure and Applied Chemistry (IUPAC) in May 2012, after the laboratory it was named for.

PHYSICAL PROPERTIES:

Physical Property	Description
Atomic Symbol	Lv
Atomic Number	116
Atomic Mass	293 amu
State at Room Temp	Solid (metal, predicted)
Color	Unknown (likely metallic or silvery, based on similarity to polonium)
Odor	Odourless
Density	Estimated around 12.9 g/cm³
Melting Point	Unknown
Boiling Point	Unknown
Solubility in Water	Insoluble in water
Conductivity	Expected to conduct electricity and heat, like other metals
Abundance in Earth	Extremely rare, entirely synthetic

CHEMICAL PROPERTIES:

Chemical Property	Description
Reactivity	Predicted to behave similarly to polonium, forming compounds with a +2 and +4 oxidation state
Oxidation States	+2 and +4 (predicted)
Electronegativity	Estimated to be around 2.0 (Pauling scale)
Reaction with Water	Likely unreactive with water due to its short half-life
Reaction with Acids	Could form salts in strong acidic environments, but data is unavailable due to its instability
Reaction with Oxygen	May form oxides, but would decompose rapidly due to radioactivity
Flammability	Non-flammable
Toxicity	Extremely toxic and radioactive, posing severe health risks
Biological Importance	No known biological role; harmful to living organisms due to its radioactivity
Isotopes	Livermorium-293 (most stable), Livermorium-292, Livermorium-291 (used in research)

TENNESSINE

Hello pioneers! I'm Tennessine, named after Tennessee, a state with strong ties to nuclear research. I'm synthetic and radioactive, adding a new layer to the periodic table as scientists probe the heaviest elements.

INTERESTING FACTS ABOUT TENNESSINE

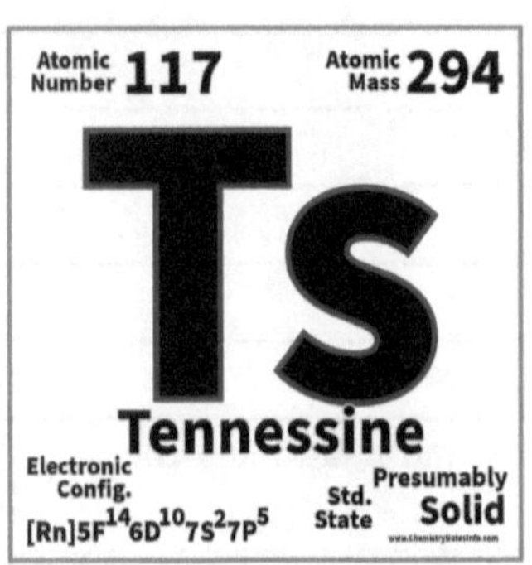

❖Tennessine is named after the U.S. state of Tennessee, where important research institutions like Oak Ridge National Laboratory and Vanderbilt University are located.

❖Tennessine was first synthesised on April 5, 2010, by a collaborative team of Russian and American scientists at Joint Institute for Nuclear Research (JINR) in Dubna, Russia.

❖Tennessine is a highly radioactive element, with its most stable isotope, T-294, having a half-life of approximately 80 milliseconds.

❖ Tennessine belongs to group 17 in periodic table, halogen group, which includes fluorine, chlorine, bromine, iodine, and astatine.

❖ Like other superheavy elements, Tennessine does not occur naturally and is synthesised in a laboratory.

❖ Tennessine is produced by bombarding berkelium-249 with calcium-48 ions in a particle accelerator, leading to the fusion of the two elements.

❖ Only a few atoms of Tennessine have ever been synthesised, making its properties difficult to study and understand.

❖ Unlike other halogens, Tennessine is predicted to have metallic properties, although its chemical behaviour is still not fully known.

❖ Tennessine was officially named by the International Union of Pure and Applied Chemistry (IUPAC) in November 2016, in recognition of Tennessee's contribution to nuclear research.

❖ Due to its short half-life and high radioactivity, Tennessine has no practical applications & is used for research into superheavy elements.

❖ Tennessine is heaviest member of halogen group and is expected to behave differently from lighter halogens like iodine and astatine.

PHYSICAL PROPERTIES:

Physical Property	Description
Atomic Symbol	Ts
Atomic Number	117
Atomic Mass	294 amu
State at Room Temp	Solid (metallic, predicted)
Color	Unknown (likely metallic, possibly silvery or grey)
Odor	Odourless
Density	Estimated to be very high, similar to other superheavy elements
Melting Point	Unknown
Boiling Point	Unknown
Solubility in Water	Insoluble in water
Conductivity	Expected to conduct electricity and heat, potentially a good conductor
Abundance in Earth	Extremely rare, entirely synthetic

CHEMICAL PROPERTIES:

Chemical Property	Description
Reactivity	Expected to be less reactive than iodine and astatine, but may form compounds with oxidation states of -1, +1, +3, and +5
Oxidation States	-1, +1, +3, +5 (predicted)
Electronegativity	Estimated to be around 1.7 (Pauling scale)
Reaction with Water	Likely to be unreactive with water
Reaction with Acids	Predicted to form salts in acidic environments
Reaction with Oxygen	May form oxides, though decomposition would occur due to its radioactivity
Flammability	Non-flammable
Toxicity	Extremely toxic and radioactive, posing severe health risks
Biological Importance	No known biological role; harmful to living organisms due to its radioactivity
Isotopes	Tennessine-294 (most stable), Tennessine-293, Tennessine-292 (used in research)

OGANESSON

Hey boundary pushers! I'm Oganesson, the heaviest known element. My atomic structure is still a bit of a mystery, and I'm highly radioactive, existing for mere moments before decaying. I represent the outermost edge of the periodic table!

INTERESTING FACTS ABOUT OGANESSON

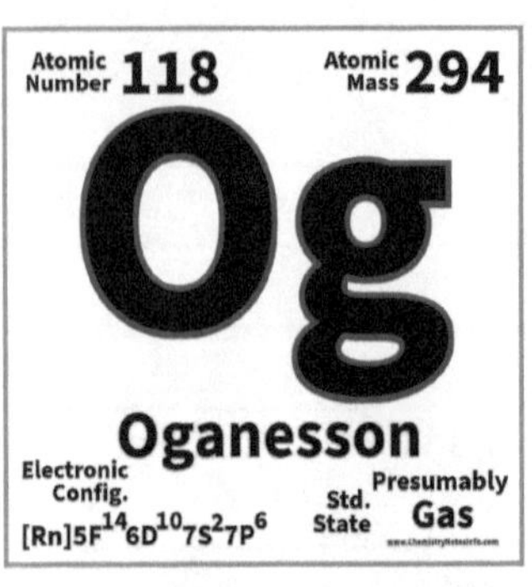

❖Oganesson is named in honour of Yuri Oganessian, a prominent Russian nuclear physicist known for his contributions to the discovery of superheavy elements.

❖Oganesson was first synthesised in 2002 by a collaborative team of Russian and American scientists at the Joint Institute for Nuclear Research (JINR) in Dubna, Russia.

❖Oganesson is a highly radioactive element, with its most stable isotope, Oganesson-294, having a half-life about 0.89 milliseconds.

❖ Oganesson is not found in nature and can only be produced in laboratories through nuclear reactions.

❖ Oganesson belongs to group 18 of the periodic table, noble gases, which also includes helium, neon, argon, krypton, xenon, & radon.

❖ Despite being in the noble gas group, Oganesson is predicted to behave more like a metalloid or even a solid rather than a gas due to relativistic effects on its electrons.

❖ Only a few atoms of Oganesson have been synthesised, making it extremely difficult to study in detail.

❖ Oganesson is produced by bombarding californium-249 with calcium-48 ions in a particle accelerator, leading to the fusion of these elements.

❖ Oganesson is currently the heaviest known element, with an atomic number of 118.

❖ Oganesson was officially named by the International Union of Pure and Applied Chemistry (IUPAC) in November 2016, in recognition of Yuri Oganessian's contributions to nuclear science.

❖ Oganesson's nucleus is highly unstable, and it quickly decays into lighter elements through alpha decay, releasing energy in process.

PHYSICAL PROPERTIES:

Physical Property	Description
Atomic Symbol	Og
Atomic Number	118
Atomic Mass	294 amu
State at Room Temp	Solid (predicted, although noble gases are typically gaseous, Oganesson is expected to behave differently)
Color	Unknown (likely metallic or grey due to its predicted solid state)
Odor	Odourless
Density	Estimated extremely high, possibly over $10g/cm^3$
Melting Point	Unknown
Boiling Point	Unknown
Solubility in Water	Insoluble in water
Conductivity	Expected to conduct electricity and heat if in solid form
Abundance in Earth	Extremely rare, entirely synthetic

CHEMICAL PROPERTIES:

Chemical Property	Description
Reactivity	Predicted to be more reactive than other noble gases due to relativistic effects on its electrons
Oxidation States	+2, +4 (predicted)
Electronegativity	Estimated to be around 2.2 (Pauling scale)
Reaction with Water	Expected to be unreactive with water due to its short half-life
Reaction with Acids	Likely to form weak bonds, but research is limited
Reaction with Oxygen	May form oxides, though it would decompose quickly
Flammability	Non-flammable
Toxicity	Extremely toxic and radioactive, posing severe health risks
Biological Importance	No known biological role; harmful to living organisms due to its radioactivity
Isotopes	Oganesson-294 (most stable), Oganesson-293 (used in research)

Periodic Table of Elements

www.ChemistryNotesInfo.com

1																	2
H Hydrogen																	**He** Helium
3 **Li** Lithium	4 **Be** Beryllium											5 **B** Boron	6 **C** Carbon	7 **N** Nitrogen	8 **O** Oxygen	9 **F** Fluorine	10 **Ne** Neon
11 **Na** Sodium	12 **Mg** Magnesium											13 **Al** Aluminum	14 **Si** Silicon	15 **P** Phosphorus	16 **S** Sulfur	17 **Cl** Chlorine	18 **Ar** Argon
19 **K** Potassium	20 **Ca** Calcium	21 **Sc** Scandium	22 **Ti** Titanium	23 **V** Vanadium	24 **Cr** Chromium	25 **Mn** Manganese	26 **Fe** Iron	27 **Co** Cobalt	28 **Ni** Nickel	29 **Cu** Copper	30 **Zn** Zinc	31 **Ga** Gallium	32 **Ge** Germanium	33 **As** Arsenic	34 **Se** Selenium	35 **Br** Bromine	36 **Kr** Krypton
37 **Rb** Rubidium	38 **Sr** Strontium	39 **Y** Yttrium	40 **Zr** Zirconium	41 **Nb** Niobium	42 **Mo** Molybdenum	43 **Tc** Technetium	44 **Ru** Ruthenium	45 **Rh** Rhodium	46 **Pd** Palladium	47 **Ag** Silver	48 **Cd** Cadmium	49 **In** Indium	50 **Sn** Tin	51 **Sb** Antimony	52 **Te** Tellurium	53 **I** Iodine	54 **Xe** Xenon
55 **Cs** Cesium	56 **Ba** Barium		72 **Hf** Hafnium	73 **Ta** Tantalum	74 **W** Tungsten	75 **Re** Rhenium	76 **Os** Osmium	77 **Ir** Iridium	78 **Pt** Platinum	79 **Au** Gold	80 **Hg** Mercury	81 **Tl** Thallium	82 **Pb** Lead	83 **Bi** Bismuth	84 **Po** Polonium	85 **At** Astatine	86 **Rn** Radon
87 **Fr** Francium	88 **Ra** Radium		104 **Rf** Rutherfordium	105 **Db** Dubnium	106 **Sg** Seaborgium	107 **Bh** Bohrium	108 **Hs** Hassium	109 **Mt** Meitnerium	110 **Ds** Darmstadtium	111 **Rg** Roentgenium	112 **Cn** Copernicium	113 **Nh** Nihonium	114 **Fl** Flerovium	115 **Mc** Moscovium	116 **Lv** Livermorium	117 **Ts** Tennessine	118 **Og** Oganesson

57 **La** Lanthanum	58 **Ce** Cerium	59 **Pr** Praseodymium	60 **Nd** Neodymium	61 **Pm** Promethium	62 **Sm** Samarium	63 **Eu** Europium	64 **Gd** Gadolinium	65 **Tb** Terbium	66 **Dy** Dysprosium	67 **Ho** Holmium	68 **Er** Erbium	69 **Tm** Thulium	70 **Yb** Ytterbium	71 **Lu** Lutetium
89 **Ac** Actinium	90 **Th** Thorium	91 **Pa** Protactinium	92 **U** Uranium	93 **Np** Neptunium	94 **Pu** Plutonium	95 **Am** Americium	96 **Cm** Curium	97 **Bk** Berkelium	98 **Cf** Californium	99 **Es** Einsteinium	100 **Fm** Fermium	101 **Md** Mendelevium	102 **No** Nobelium	103 **Lr** Lawrencium

Congratulations! Now you know lots of facts about all elements of periodic table.

Still curious to know much more, visit our science website www.ChemistryNotesInfo.com To get our science books, visit at Amazon, Flipkart or NotionPress website/app and search author name 'Jitendra Singh Sandhu' Or scan this QR code.

✒ MY SIENCE THOUGHTS

This space is just for you — a place where you can bring your scientific ideas to life on paper

____

____

____

____

____

____

____

____

____

____

____

____

____